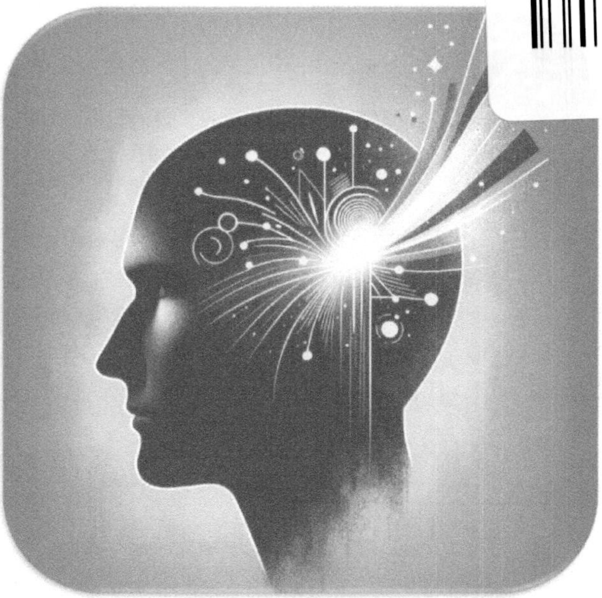

Transformation Through Curiosity

The Little Book of Chats

A Practical Guide to

Generative AI and ChatGPT

Peter Stanford

In Collaboration with ChatGPT

First published in 2023 by Peter Stanford

All enquiries should be made to the author:
peter.stanford@mentormeaus.com

A catalogue entry for this book is available from the National Library of Australia.

ISBN: 978-0-6459261-2-5

Disclaimer

The material in this publication is of the nature of general comment only and does not represent professional advice. It is not intended to provide specific guidance for particular circumstances, and it should not be relied on as the basis for any decision to take action or not take action on any matter which it covers. Readers should obtain professional advice, where appropriate, before making any such decision. To the maximum extent permitted by law, the author and publisher disclaim all responsibility and liability to any person, arising directly or indirectly from any person taking or not taking action based on the information in this publication.

Cover design and interior design:

Peter Stanford https://www.mentormeaus.com

Book Adviser: Jaqui Lane www.thebookadviser.com.au

The paper this book is printed on is certified as environmentally friendly.

Contents

Foreword

Don't confuse the content of this book, Artificial Intelligence, with the author Peter Stanford's natural intelligence.

I met Peter in 1988 when he was appointed country manager for Novell Networks, one of the world's hottest technology companies. Peter's initial task was appointing a second Australian distributor for the rapidly growing company. Twenty-seven companies, including my tiny company, Com Tech Communications, applied for distribution rights. At the time, we had three people in Com Tech. Peter had much easier decisions than appointing Com Tech but backed his judgement and career by making what he thought was the right decision, not necessarily the easiest one. I'm proud that we never let Peter or Novell down.

That was the luckiest break that I have had in my life. I learnt a lot from Peter during the time that we partnered together – none more than "sometimes the right decisions are not necessarily the easiest ones". This mantra has served me well in my own business career.

With a career spanning over 55 years, Peter has been around for some of the most disruptive eras in technology, including the advent of the personal computer, the power of the internet, the move to mobile computing, the migration from on-premises computing to the cloud and now the next big wave – Artificial Intelligence (AI). As with any transformative technology, AI can have both advantages and disadvantages.

At the ripe old age of seventy-four, Peter has recognised the phenomenal impact that Artificial Intelligence will have on our future lives. As we stand on the cusp of a new era characterised by unprecedented advancements in computing power, data availability and algorithmic sophistication, the significance of understanding Generative AI cannot be understated.

Peter has taken the time to learn and share his knowledge by writing this valuable book. It provides a simplistic way of understanding what could otherwise be a very difficult topic. Simplicity is the ultimate sophistication. *The Little Book of Chats* aims to demystify the fundamental concepts that underpin Generative AI and ChatGPT.

Whether you are a business leader or simply a curious mind and want to discover the impact of Generative AI on our future lives, this book provides a comprehensive roadmap. Combined with examples and case studies, *The Little Book of Chats* will provide the tools to understand AI's transformative power better, fostering a deeper understanding of its capabilities and limitations.

I hope that by reading Peter's *The Little Book of Chats*, you too may say, "That was the luckiest break of my life".

David Shein

Founder – Com Tech Communications

Preface

When I decided to write a book about Artificial Intelligence (AI) and ChatGPT, I was looking for assistance in the form of a collaborator or co-author. After extensive and fascinating experiences across various topics, it became apparent that the co-author could, should, and would be ChatGPT.

Co-authors give you someone to help contribute and bounce ideas off, to review the messy first and second (and third) drafts, and to encourage you to continue. This is reminiscent of my experience in 2005 when I co-authored a University textbook with Dr Joanne Curry, a then colleague and now lifelong friend. The book was titled *Practical System Development – A Project-Based Approach.* Researching, writing, editing, and publishing took about eighteen months.

Writing *The Little Book of Chats* has been a similar experience, just without the long duration. The first draft took just twelve days. I'll tell you how long it took to complete the book towards the end.

I've combined my passion for coaching, mentoring, education, and emerging technologies into a view of how AI will significantly impact these (in fact, all) industries, learning institutions, professions and our lives.

In the last few months, ChatGPT has enabled a range of additional features, including the image it generated below in response to the following prompt:

> *I want a monochrome image I can use in my book. The image should contain a photo of a person conversing with ChatGPT, where ChatGPT is a female business or executive coach or mentor.*

Figure 1 - Image generated by ChatGPT-4, 15th November 2023

This book is the product of that unique collaboration. It is evidence of the symbiotic potential between human creativity and AI's capabilities, a fusion of traditional wisdom and cutting-edge technology.

By the time you read this, further advances in AI will already have been made. You will always be playing catch-up if you don't start now.

Introduction

About this Book

Welcome to *The Little Book of Chats*, your practical guide to navigating the ever-changing landscape of Generative AI and ChatGPT.

First things first!

You won't gain much by simply reading this book. It's a practical guide, and its value is in showing you how to use ChatGPT by following the examples and then trying it out for yourself.

Whether you do it now or when you have an epiphany about the first question you want to ask, you'll need to use ChatGPT. For this book, the free version will generally be adequate. Here's how to get access.

Browse to this address: https://openai.com/chatgpt or scan the QR code:

There are several versions of ChatGPT. If I describe a capability requiring ChatGPT-4, I'll clarify that. If not explicitly noted, the prompt will work in ChatGPT-3.5. Here's what ChatGPT-4 says about these versions.

ChatGPT-3.5: This version can understand and generate text, including the use of emojis, based on a dataset that includes information up to September 2021. It does not include web browsing or image generation capabilities.

ChatGPT-4 (Paid Subscription): This version includes all the capabilities of ChatGPT-3.5 but with improved understanding and response generation. Additionally, it features web browsing and image generation, which are not available in the earlier version.

There is also the **ChatGPT-4 API,** a more technical offering from OpenAI, designed for developers and businesses to integrate advanced AI capabilities into their applications and services. Unlike the user-facing ChatGPT-4, accessed directly by individuals for various tasks, the API provides backend access to the same advanced language model. I won't be talking about or exploring the API in this book.

The version you use, however, doesn't change how you learn to write prompts; there are just some things you can't do or do as well in ChatGPT-3.5.

Document Conventions

Throughout the book, you'll see sets of prompts and responses. A prompt is indicated by a vertical line on the left side of the prompt, with the icon above it. Prompts will be in *italics*.

This is an example of prompt formatting.

A response is indicated by a vertical line on the right side of the response, with the icon above it.

This is an example of response formatting.

If I ask a question or receive a response from Bing, I'll note that. Otherwise, all prompts and responses are from ChatGPT. If I edit a response, other than changing the spelling, I'll denote it by saying **(Edited)**. Otherwise, all responses will be exactly as received from ChatGPT.

NOTE: Many responses are long (sometimes multiple pages), so I may extract a precis and provide that as the first level of response.

What is Generative AI?

Generative AI is a type of Artificial Intelligence that specialises in creating new content, be it text, images, or even code, based on the input it receives. Unlike traditional AI systems that primarily analyse or process existing data, Generative AI can produce original outputs – from writing a poem to drafting a business report. This is achieved through sophisticated algorithms that learn patterns and structures from large datasets, enabling the AI to generate coherent, contextually relevant responses often indistinguishable from human-created content. At the heart of this capability is a deep understanding of language, symbols, or images, allowing the AI to replicate existing styles and innovate within the parameters set by its programming. In essence, Generative AI is like a versatile digital artist or writer adept at crafting unique creations from the seeds of ideas provided by users.

What is ChatGPT?

ChatGPT is a state-of-the-art language model developed by OpenAI designed to generate human-like text based on the prompts it receives. It excels in a variety of tasks, including answering questions, writing essays, creating fictional stories, and even generating images using integrated tools like DALL-E. The model is trained on a vast array of internet text, allowing it to produce responses that are often insightful, relevant, and contextually appropriate.

However, there are certain limitations to what ChatGPT can do. While it can generate text and even write lyrics, it does not compose music. Its ability to create original content is grounded in the text it has been trained on, meaning it can replicate styles and formats but doesn't innovate as a human artist or creator might. ChatGPT also operates within ethical guidelines and programming constraints; it avoids generating harmful or inappropriate content and cannot engage in personal tasks like making phone calls or accessing personal data.

Moreover, ChatGPT's knowledge is limited to its training data, which only extends up to a certain point in time (last updated in April 2023). This means it may not have information on recent events or specific niche topics. Additionally, while it can simulate conversation and provide information, it lacks true understanding or consciousness, operating purely within the realm of algorithmic processing.

In summary, ChatGPT is a powerful tool for text generation and a wide range of creative tasks, but it's important to recognise its limitations and the areas where human creativity and expertise remain irreplaceable.

What is a Prompt?

> *What is your definition of a Prompt, please?*

A Prompt is an initial statement, question, or input provided by a user that sets the direction for the AI's response. It serves as the starting point for the AI's content generation, guiding the subject matter, tone, and scope of the ensuing conversation or output. Prompts can range from specific questions seeking detailed information to open-ended statements encouraging creative or analytical outputs.

In essence, a Prompt is a trigger for the AI's response process, dictating what the AI talks about or the type of content it generates. It's the primary tool through which users communicate their needs, interests, or curiosities to the AI, thereby initiating a responsive interaction. The effectiveness and relevance of the AI's response often depend on the clarity and specificity of the prompt provided by the user.

What is a Response?

What is your definition of a Response, please?

A Response is the AI's reply or output generated in reaction to a user's prompt. It encompasses the information, analysis, creative content, or guidance provided by the AI, tailored to address the user's query or input. The response is formulated based on the AI's understanding of the prompt, incorporating relevant knowledge, contextual insights, and the guidelines set forth by any custom instructions.

In essence, a Response is the AI's contribution to the conversation or interaction, reflecting its capabilities in processing language, understanding context, and generating relevant and coherent outputs. Responses can vary in length, complexity, and format, depending on the nature of the prompt and the specific requirements or expectations of the user. They are central to the interactive experience with the AI, serving as the primary means through which the AI communicates, informs, or assists the user.

What is an Interaction?

What is your definition of an Interaction, please?

An Interaction is a communicative exchange between the user and the AI, typically comprising a prompt provided by the user and the subsequent response generated by the AI. This prompt/response pair forms the fundamental unit of communication, where the user initiates the dialogue with a question, statement, or request, and the AI responds in a manner that reflects its understanding and processing of the prompt.

ChatGPT will always respond to a prompt. The AI is programmed to engage with and respond to user prompts, even if the response is simply to acknowledge or confirm something.

Every prompt elicits a response, even if that response is minimal or non-substantive. The AI's primary function is to interact and respond to user inputs, so in a way, not responding at all (i.e., complete silence) would be contrary to its design and purpose.

To summarise, this book is about using ChatGPT to write a book about ChatGPT (excuse the tautology), so it's a combination of a primer on ChatGPT and examples of use cases to get you started. Later, I'll provide more detailed examples of how to maximise and master your interaction with ChatGPT.

The Exponential Acceleration of AI

Artificial Intelligence (AI) is rapidly evolving, and it's easy to feel overwhelmed by the sheer volume of information. This book will cut through the noise, offering practical examples and information that can be readily applied. Whether you're an executive, a leader, or a curious mind, these pages are designed to be digestible, actionable, and, most importantly, accessible - no PhD required. Just as the best conversations are the ones that leave you enlightened, not exhausted, consider this your guide to meaningful chats with AI.

If you're wondering what gives me the gall to write such a book, let me offer you a snapshot of my journey.

With over fifty-five years in the business world, I've seen trends come and go, but the potential of AI has been a game-changer. I've worn many hats: CEO, coach, mentor, business owner, and university lecturer.

I hold a Master's degree in Software Engineering and a Graduate Certificate in Applied Coaching, credentials that have fuelled my curiousity and passion for effective communication and leadership. But credentials aside, perhaps my most daring feat has been a career pivot at age seventy-two, diving headfirst into my Coaching and Mentoring practice (go to www.mentormeaus.com or scan the QR code below to visit my website).

Why this book? I've had the privilege of engaging in over 200 chats (and counting) with ChatGPT, exploring its capabilities and limitations while blending its generative prowess into my coaching and mentoring practice. This book isn't just a tech manual; it combines traditional coaching, mentoring, teaching methods, and AI-powered insights. I want to share with you the technical know-how and the nuanced human aspect of using AI as a tool for personal and professional growth.

Staying Relevant in a Rapidly Evolving AI World

Creating relevant and valuable content over time is vital in the rapidly evolving landscape of AI and tools like ChatGPT. This concept, known as "evergreen" content, is especially pertinent in a field marked by constant change and innovation. The core focus of this book on teaching you about prompts, responses, and custom instructions adheres to this principle.

While the specific functionalities and capabilities of AI models like ChatGPT may advance, the fundamental skills of crafting effective prompts, interpreting responses, and utilising custom instructions will retain their importance. These skills are the bedrock of effective AI interaction, transcending the specifics of any single version or iteration of the technology.

Understanding how to communicate with AI, extract the most value from its responses, and tailor its functionality to specific needs is a timeless skill set. It's akin to learning the principles of a language rather than memorising a set of phrases; the former remains useful regardless of changes in vocabulary or dialect.

Therefore, the insights and skills imparted in this book are designed to be as evergreen as possible, providing a lasting foundation for engaging with AI. This approach ensures that you are equipped not just for today's AI but are also prepared to adapt to the AI of tomorrow.

Who is this Book for?

When I began my studies in Applied Coaching in September 2021, I became increasingly curious about how technology would impact the fields of Coaching, Mentoring, and Education. I'm not particularly academic, but I love learning. The pursuit of knowledge is more important to me than a qualification.

I am also impatient. I want that knowledge as soon as possible.

While studying, I purchased various books on Coaching in addition to the prescribed texts. Of particular interest were books on:

- Cognitive Psychology
- Self-Determination Theory
- Emotional Intelligence
- Coach the Person, Not the Problem
- The Psychology of Coaching and Mentoring

ChatGPT has subsequently enabled me to research the essence of all these books in a fraction of the time with a level of trust and reliability that allows me to form an opinion that I am comfortable talking about and sharing.

There isn't a knowledge worker function, job role or activity that AI won't impact. Despite their resistance, universities, schools, and other educational institutions will also be impacted. AI will touch everyone in some way, shape, or form.

Whether you're a tech-savvy guru or an AI novice, I invite you to join me on this exciting journey. Let's explore what it truly means to integrate AI into our lives and how we can leverage it to become more insightful, resilient, productive, and effective in our respective fields.

Whoever you are, whatever role, position or title you hold, open your mind to the possibilities and opportunities that Generative AI and ChatGPT bring.

Who (or What) Really Wrote this Book?

If you know anything about AI and ChatGPT (don't worry, I'll enlighten you if you don't), there's more to it than saying: "Write me a book about AI and ChatGPT".

ChatGPT didn't generate or write any of the prompts by itself. I'll demonstrate later how I used ChatGPT to refine those prompts and improve the quality and context of the question, but all the prompts originated from me. I also use Grammarly to check my spelling and grammar as I type, whether in any of the Microsoft 365 apps or inside a browser. So, when I say, "I wrote the prompts", those prompts might also have been edited before submission.

Responses are not as straightforward. Yes, fundamentally, each response you see in the book has been generated by ChatGPT, but I may have altered it for readability and style or to augment or supplement the response with an opinion or insight of my own.

I do this because the book isn't going to be read by AI; it's going to be read by humans. If it's not interesting and informative, you won't read it. I'm also aware of the variations between how ChatGPT generates responses and how other programs interpret the spelling and grammar of that output.

So, what percentage of the book was written by me, and what percentage by ChatGPT? Is the balance in favour of me or ChatGPT? You be the judge.

Who Owns the Copyright?

This was an important question when I began to write. Who would own the material in the book, understanding that it was a combination of my words, my input in the form of prompts, and the responses from ChatGPT? To show how omnipresent ChatGPT is, I created the following prompt in Bing.

(Bing)

> *Considering the issue of plagiarism and the recent announcement by OpenAI and Microsoft about covering the costs of people or organisations accused of breaching copyright, am I correct in assuming that your training data was in the public domain and, therefore, not likely to be challenged? I appreciate that as a Generative AI, you don't extract content verbatim or claim it as your own.*

Bing provided a comprehensive response, including links to external sources. However, I wanted a shorter response. I asked ChatGPT the following question (I could as easily have asked Bing, too). This demonstrates the seamless integration between the two, remembering that ChatGPT cannot access what I asked Bing and vice versa.

> *I know Bing is now powered by ChatGPT-4, and I've used both to ask questions to discuss their relative merits in my book. I want to include a paragraph about the Copyright Shield and the ownership of content produced by ChatGPT, etc. The response from Bing was very detailed; I'd like a 150-200-word summary, please.*

When using AI platforms like Bing, powered by ChatGPT-4, or ChatGPT itself, it's important to understand the nuances of copyright and content ownership. The AI generates responses based on a vast corpus of pre-existing information, including publicly available texts, books, websites, and other media. While the AI's responses are original compositions, they can occasionally echo or resemble the source material. This raises questions about copyright, especially in contexts where the AI's output is used commercially or published. To address this, Bing employs a 'Copyright Shield' – a mechanism designed to reduce the likelihood of generating content that infringes on copyright. This shield does not guarantee complete avoidance of copyright issues but serves as a proactive step. Users of AI-generated content should be aware that while they hold the rights to the AI-produced text, this does not exempt them from copyright considerations, particularly when the content closely mirrors protected works. It's a complex intersection of AI innovation and copyright law, emphasising the need for cautious and informed use of these technologies.

Venturing into Generative AI and ChatGPT may seem like stepping into an intricate labyrinth of algorithms, Large Language Models (LLMs), and decision trees. It is easy to get overwhelmed by the jargon. However, rest assured, this book is designed with you, the real-world user, in mind.

While I'll provide some definitions and context to help paint a clearer picture, the heart of this book isn't about getting bogged down in technicalities. Instead, it is about harnessing the immense power of these tools, understanding how to frame your questions, and getting the best possible responses. In essence, it is about conversation, not computation.

So, as you flip through these pages, remember that you do not need a degree in computer science or a background in Artificial Intelligence. All you need is curiosity and a desire to engage in meaningful dialogue with the future of technology (*See Appendix C – Demystifying the Jargon*).

While these terms are foundational to understanding the mechanics behind AI, I will focus more on practical interactions and real-world applications for our journey through this book. So, for now, you can place these terms in your mental back pocket and enjoy the conversation.

Figure 2 - Image generated by ChatGPT-4, 27th November 2023

DALL-E Image Description

An abstract monochrome image representing the concept of context. The design features a central point of focus, surrounded by thin, delicate, interconnected lines arranged to create a sense of depth and complexity.

CHAPTER 1 – It's All About The Context

Context is Everything!

In any conversation, the depth and relevance of the answers are most often determined by the clarity and detail of the question. When engaging with Generative AI, this becomes even more pronounced. The system doesn't have memories or inherent knowledge about the individual user but rather responds based on the information provided at that moment. By furnishing the AI with ample context, you're guiding it, offering a clearer picture of what you seek.

Imagine you're in a vast library. Without specifying which book or topic you're interested in, asking a librarian for 'a good book' is likely to yield a broad recommendation or, more likely, a question from the librarian about your interests. But if you specify, for instance, that you want a 'thriller set in 19th century London' or want to know what industry experts like Mo Gawdat are saying about AI, the recommendation becomes more tailored.

Similarly, giving ChatGPT more context narrows its vast knowledge to cater to the most relevant information to respond to your query. This means your questions can elicit more accurate, insightful, and useful answers. The mantra here? The clearer the context, the better the response.

What is a Custom Instruction?

> *What is your definition of Custom Instructions, please?*

Custom Instructions are specific directives or guidelines provided by a user to tailor the AI's responses to their unique preferences or requirements. These instructions can dictate the style, structure, format, or content of the AI's replies. They serve to personalise the interaction, ensuring that the responses align more closely with the user's expectations, needs, or the context of their inquiry. Custom Instructions can be established as ongoing preferences for all interactions or be specified on a case-by-case basis within individual prompts.

In practice, Custom Instructions enable users to shape the conversation in a way that best suits their purpose, whether it's for coaching, research, learning, or casual conversation. This customisation enhances the relevance and usefulness of the AI's responses, making the interaction more effective and efficient for the user.

You're not locked into this initial information. As you'll see in the next few pages, you can change these anytime.

Custom Instructions (Contextual)

There is a lot of flexibility in the way you interact with ChatGPT. You can create a prompt that asks for something specific that only applies to the current chat. You create custom instructions if you want ChatGPT to know things about you and respond in certain ways across all your chats. These can be both static (what you might call general guidelines) and dynamic (changing per interaction based on immediate needs or preferences).

I recommend creating these contextual custom instructions when you first start using ChatGPT. This provides context for the way you interact with ChatGPT.

To access your custom instructions, click your name at the bottom left side of the chat window. It will look something like this:

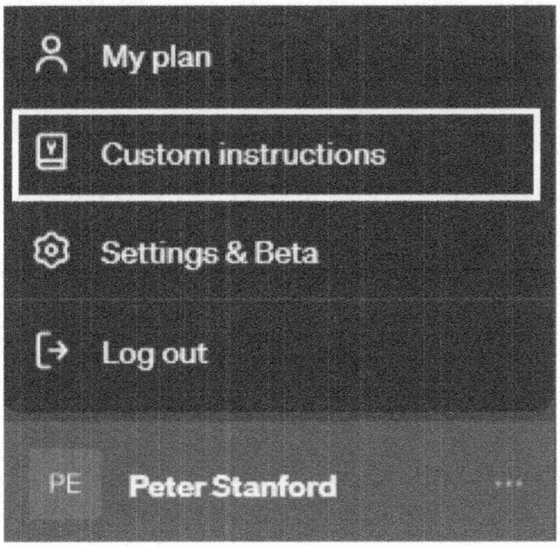

Click the 'Custom instructions' menu option. This opens a window into which you can enter custom instructions that apply to all new chats. Note the caveat 'new' here. Whatever context you provide for these custom instructions only applies to chats you create, not ones you've already created. There are two text boxes into which you can enter content.

The first is titled "**What would you like ChatGPT to know about you to provide better responses?**".

You have 1,500 characters to describe what ChatGPT wants to know about you. A set of thought starters appears when this text box has the focus.

Thought starters

- Where are you based?
- What do you do for work?
- What are your hobbies and interests?
- What subjects can you talk about for hours?
- What are some goals you have?

You decide what you want to say. You're not limited to the thought starters displayed; however, these will give ChatGPT the context to understand who you are when you ask a question.

The second is titled "**How would you like ChatGPT to respond?**".

A set of thought starters also appears when this text box has the focus.

Thought starters

- How formal or casual should ChatGPT be?
- How long or short should responses generally be?
- How do you want to be addressed?
- Should ChatGPT have opinions on topics or remain neutral?

For transparency, these are the custom instructions from my ChatGPT profile. I've tried giving ChatGPT as much information about me as possible to improve its understanding of how I create prompts and what I expect from them in return.

These are normally private to you. I've shared them here to show what you might want to include in your custom instructions.

About Me

As an Australian business and executive coach with over five decades of leadership experience, including CEO roles, I specialise in integrating traditional coaching techniques with cutting-edge Generative AI. This unique blend aids in fostering self-awareness and driving behavioural change among my clients. My approach involves meticulously crafting prompts to elicit profound insights. My diverse background, including performing arts and sports achievements, has enriched my coaching philosophy. My journey as a NSW State Junior Relay Champion and a hockey representative, coupled with the resilience developed from overcoming a career-impacting back injury, deeply informs my perspectives on goal-setting and endurance. My goal is to pioneer the use of AI in executive coaching, striving for excellence and recognition in this innovative field.

Similarly, I've told ChatGPT what I expect in return, i.e., how I want ChatGPT to respond. I'll show how I've amended this over time.

About the Response

When wrapping up a chat session, I'll use the emoji 🙌. In response, I'd like a 150-word prompt summary, a 200-word response summary, and a 150-word précis outlining the chat's overall context. Specific emoji like 👨‍🎓 and 🎇 will signal that I'd like to see a numbered list of the prompts and responses, respectively. My responses should strike a balance between informality and detail, cutting to the chase while avoiding any fluff. The default dialect should be Australian English; I'd appreciate source citations for credibility whenever possible. You can refer to me as 'Coach' or 'Peter.' Lastly, I value ChatGPT's

> opinions, as they contribute to my ongoing research and facilitate deeper, more meaningful conversations in the future.

In the earlier explanation of creating custom instructions, you saw the words I used when I created prompts. You'll notice several emojis in the section about how I want ChatGPT to respond.

These emojis act as shortcuts, just like key combinations in most applications you use, for example, CTRL/A, CTRL/C, CTRL/V, and CTRL/X. I'm not an Apple Mac user, so I can't tell you the equivalent key combinations, but I've shared a prompt on the next page that you can use to find out.

Using emojis allows me to enter a single character into a prompt rather than typing a custom instruction. Because I've configured them in the global custom instructions, they will be activated in every new chat I create.

Emojis are a relatively new feature. On a Windows device, they can be accessed via the key combination of the Windows key plus a decimal point (**Windows + .**). This pops up a window, giving you an extensive choice of emojis.

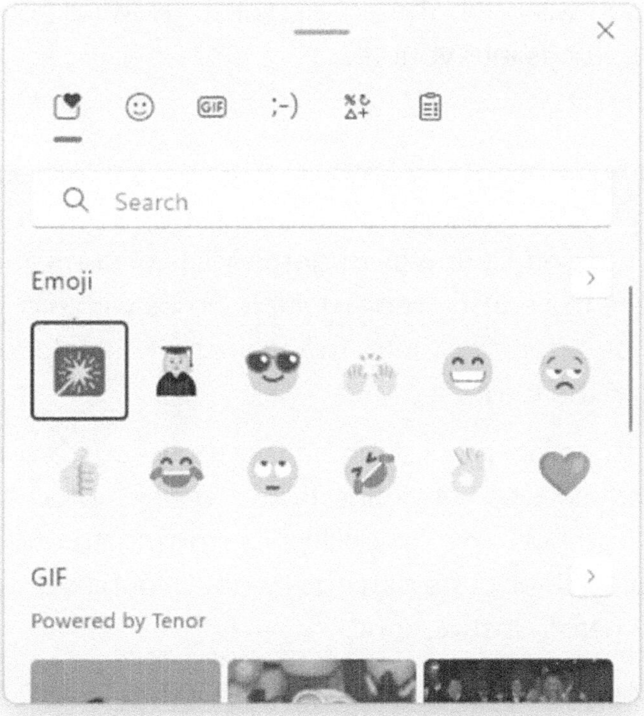

Are you using an Apple device? To find out what the key combination is, create a prompt asking something like:

What is the key combination for displaying the emoji popup on an Apple device?

You'll get a response similar to the following:

On an Apple device, such as a Mac, you can bring up the emoji popup by pressing **Control + Command + Space**. This key combination will open the emoji picker, allowing you to select and insert emojis into your text.

I could have used any emoji in my custom instructions; there wasn't a particular reason for choosing the ones I did, but they provided a more efficient way of getting lists or summaries within a chat. I can use the same emoji multiple times within a chat, i.e., ask for a precis at different points. The last one will include everything preceding it in the chat.

It's worth some time deciphering these custom instructions, particularly those where I define how I want ChatGPT to respond. First, I want to show my amended custom instructions. As I use ChatGPT more, I think of new things I want to achieve, which means altering my custom instructions.

Can you spot the differences between the two versions?

About the Response

Please display 'Last Modified Date: ' followed by the date in DD/MM/YYYY format at the beginning of each response.

When wrapping up a chat session, I'll use the emoji 🙌. In response, I'd like a 150-word prompt summary, a 200-word response summary, and a 150-word précis outlining the chat's overall context. Specific emoji like 🧑‍🎓 and 🎇 will signal that I'd like to see a numbered list of the prompts and responses, respectively. My responses should strike a balance between informality and detail, cutting to the chase while avoiding any fluff. The default dialect should be Australian English, and whenever possible, I'd appreciate source citations for the sake of credibility. You're welcome to refer to me as 'Coach' or 'Peter.'

Lastly, I value ChatGPT's opinions, as they contribute to my ongoing research and facilitate deeper, more meaningful conversations in the future.

One of the many challenges with ChatGPT is that it doesn't keep metadata about individual chats, for example, information about the date and time a chat was created or modified. This may not be a problem if you create simple chats, execute them on a single day, and then put them away.

I interact with chats repeatedly, augmenting or revising them based on what I'm thinking or researching. I want to track when I start, modify, edit or update each chat.

Another challenge is that custom instructions only allow 1,500 characters. I added another instruction to display the date I started the chat, and then I wanted to know how many days I'd been working on that chat.

Please display 'Last Modified Date: ' followed by the date in DD/MM/YYYY format at the beginning of each response. Also, use the 'Last Modified Date' from the first response as the start date to calculate and display the duration of the chat by stating, 'You have been working on this chat for 'n' days'. When calculating the duration of the chat, count each calendar date on which the chat was active as one full day. This means if the chat was active on different dates, irrespective of the time spent on each day, each date should be counted as a separate day.

When wrapping up a chat session, I'll use the emoji 🙌. In response, I'd like a 150-word prompt summary, a 200-word response summary, and a 150-word précis outlining the chat's overall context. Specific emoji like 🧑‍🎓 and 🎇 will signal that I'd

> *like to see a numbered list of the prompts and responses, respectively. My responses should strike a balance between informality and detail, cutting to the chase while avoiding any fluff. The default dialect should be Australian English, and whenever possible, I'd appreciate source citations for the sake of credibility. You're welcome to refer to me as 'Coach' or 'Peter.' Lastly, I value ChatGPT's opinions, as they contribute to my ongoing research and facilitate deeper, more meaningful conversations in the future.*

This used 1,332 of the 1,500-character limit, making adding more features to my custom instructions difficult. I needed to improve my custom instruction to enable further configuration of how I want ChatGPT to respond.

Who better to ask than ChatGPT to provide a shorter version that didn't impact what I wanted to achieve?

I gave it this prompt.

> *Please let me give you the custom instructions I've created for responses and see whether you can improve them without diminishing the functionality.*

The response reduced the character count from 1,332 to 586, which now looks like this.

Please begin each response with 'Last Modified Date: [DD/MM/YYYY]'. Calculate and display the chat's duration in days at the start, considering each active date as a full day.

At chat wrap-up (indicated by 🙌), provide a 150-word summary of prompts, a 200-word response summary, and a 150-word overall context précis. Use 👨‍🎓 and 🎇 emojis to request numbered lists of prompts and responses, respectively.

Responses should be concise, detailed, and in Australian English. Include source citations for credibility. Refer to me as 'Coach' or 'Peter.' Your opinions are valued for my research.

Testing changes to custom instructions means creating a new chat each time. I won't go through every iteration, but I progressively refined the custom instruction until I landed on this.

Please begin each response with 'Last Modified Date: [DD/MM/YYYY] (Active Days: n)', where n represents the count of the unique dates on which the chat was active. Calculate and display the number of unique dates on which the chat was active, counting each date only once.

At chat wrap-up (indicated by 🙌), provide a 150-word summary of prompts, a 200-word response summary, and a 150-word overall context précis. Use 👨‍🎓 and 🎇 emojis to request numbered lists of prompts and responses, respectively.

Responses should be concise, detailed, and in Australian English. Include source citations for credibility. Refer to me as 'Coach' or 'Peter.' Your opinions are valued for my research.

To prove this works, I had to create a new chat and wait at least one day before accessing it again. The days don't matter, but if I create the chat on a Sunday, I can't access it again until Tuesday. Here's an example of what the chat response would be.

Last Modified Date: 19/11/2023 (Active Days: 1)

Etc.

Getting involved with custom instructions to this level of detail may not be important for you. However, you should now understand how to create, update, and test them.

In summary, while it may have taken a while, you've seen how flexible, adaptable, and powerful the custom instruction feature in ChatGPT is. When you interact with ChatGPT and treat it as the expert in the room, you'll always get more specific answers to your questions.

Here's the version of the custom instructions at the time of publishing. I'm sure I'll change it again over time.

Please initiate each response with 'Last Modified Date: [DD/MM/YYYY] (Active Days: n)', where 'n' represents the count of distinct active chat dates. At the end of our conversation (indicated by 🙌), provide a 150-word summary of prompts, a 200-word synthesis of responses, and a 150-word overall context précis. Use 🧑‍🎓 and 🎇 emojis for numbered lists of prompts and responses, respectively. For concise and detailed responses, adhere to Australian English and include source citations for credibility. Address me as 'Coach' in shared chats for privacy.

Upon using the 📄 emoji, generate a table with the following columns: Chat Title, Creation Date, Precis, and Hashtag Suggestions. Use the words immediately following the 📄 emoji in the chat as the 'Chat Title'. The 'Creation Date' should be formatted as DD/MM/YYYY. The 'Precis' should be a 100–150-word summary of the chat's content, and include two or three relevant 'Hashtag Suggestions' that encapsulate key themes or topics discussed.

Custom Instructions (Interaction-specific)

When you override a standing custom instruction within a specific interaction, that override itself becomes a temporary custom instruction for that particular interaction. For example, if I ask for a 50-word précis instead of the standard 150-200-word précis I set as a general guideline, this is effectively a custom instruction for that specific prompt.

A simple example is when you give ChatGPT a prompt that says something like:

Please provide a precis of 150-200 words of the chat to this point.

The instruction provides a clear directive to ChatGPT. The word count range allows ChatGPT some flexibility. The response will generally be within this range (which is purely arbitrary). You might want more or less, so change it when you give ChatGPT the prompt.

You could write this prompt differently but still get the same response. For example:

> *Provide a 150–200-word precis.*

ChatGPT understands both variants. By definition, wherever you are in the chat, the custom instruction will generate the precis to that point. However, you might only want a precis for certain chat interactions.

In this context, an interaction is a single prompt/response pair.

Let's consider that you want a precis of selected elements. Your prompt could be something like this.

> *Can you provide a 150-200-word precis of the first and fifth interactions only?*

If you have at least five prompts in your chat, this will work as expected; however, what does ChatGPT do if your prompt is invalid? Let's consider this.

> *Please provide a 150-200-word precis of the first and **fiftieth** interactions and combine them into a single precis.*

I've highlighted the word **fiftieth** to make the error obvious. Here's a snippet of how ChatGPT responded.

Since our conversation has not extended to a fiftieth interaction, I'm unable to provide a precis for it. However, I can still offer a combined precis of the first interaction and a general overview of our conversation thus far, within the requested word limit:

Cont'd . . .

Next, I wanted to show the response if you typed the single word **Precis** into the prompt. Remember that before I did this, I had given ChatGPT an erroneous instruction, and before that, I had asked for a precis of the first and fifth interactions only.

Using a single-word prompt, I asked for a precis.

Precis

ChatGPT responded with a precis of the first and fifth interactions because that was the last valid context.

Precis of chat

This time, the response was of the entire chat. Interestingly, it was about 148 words, i.e., only marginally on the low side of the 150-200-word request.

The takeaway is how ChatGPT remembers the context within a chat but not across chats.

How Does ChatGPT Interpret Your Custom Instructions?

How do you know how ChatGPT applies your custom instructions when it responds to your prompts? The best way to find out is to ask it. Here's an example of how you might do that.

> *I want you to look at the first part of my custom instruction and outline what it tells you about me and how it might influence how you respond.*

Copy and paste the content, and see how ChatGPT responds. If you don't like what ChatGPT tells you, there are a few options:

1. Regenerate the response
2. Read what ChatGPT said and critically question whether your custom instruction really addresses who you are.

In summary, you've seen how to 'configure' ChatGPT to have it interact with you meaningfully.

Striking the Balance

When providing context for ChatGPT, think of Goldilocks and the Three Bears. Offer too little, and you might find the answer too cold - vague and generalised, not quite hitting the mark. Drown the model in detail, and the response could be too hot, overwhelmed by the specifics, missing out on the nuance.

The key is to strike that "just right" balance. Give enough detail to frame your question and guide the direction, but allow room for the model to manoeuvre, think (so to speak), and craft a nuanced answer. As with any tool, the input quality significantly influences the output quality. Mastering the balance between brevity and verbosity will streamline your interactions and elevate the responses' richness.

Remember, it's not about overloading or underfeeding the system; it's about serving the optimal portion to get the desired outcome.

Don't Expect Perfection

Technical Limitations

ChatGPT, like many advanced AI models, functions within a defined 'token limit,' which essentially acts as a cap on the amount of text it can process (read and write) in a single interaction. For the GPT-4 model, this limit is around 4096 tokens. A token can be as small as a word piece or as large as a word. Large words This limitation means that the content must be divided into manageable segments for comprehensive tasks requiring extensive analysis or detailed exploration of large documents.

NOTE: There is no direct correlation between a token and a word. Tokens are part of the internal architecture of Generative AI tools like ChatGPT. The token count doesn't limit the number of words you can have in a single chat; it simply means how much of that chat it can accommodate when responding to a prompt. I have one chat consisting of more than 11,000 words.

When dealing with lengthy documents or in-depth analyses that span multiple pages, it's essential to break down the material into smaller parts. Each part should be within the token limit for ChatGPT to process effectively. Then, these parts can be sequentially input into the system for analysis or response. This approach ensures the AI can handle the information efficiently without exceeding its processing capacity. However, it's important to note that the context may be limited to the content within the current token window, which could impact the continuity and depth of analysis in some cases.

This token limit is a significant consideration when using ChatGPT for complex, multi-faceted tasks, requiring strategic planning of inputs to optimise the output quality and relevance.

The Realities and Limitations of Business Writing

A common misconception about using Generative AI, like ChatGPT, for business writing is anticipating immediate perfection. While ChatGPT is an efficient and versatile tool capable of aiding in various tasks — from composing emails to creating reports — it's important to recognise its inherent limitations. The AI relies on pre-existing data and patterns to generate responses, so it may not always grasp the nuances of specific business contexts or reflect the latest industry developments. Furthermore, its output requires careful review, particularly for accuracy and relevance, as it can sometimes produce generic or overly broad content.

ChatGPT's training data limits its responses, extending only to a certain point. This can pose challenges in rapidly evolving fields or when dealing with recent information. Moreover, in business writing, where tone and style can be as crucial as content, ChatGPT may not always align perfectly with you or a company's unique brand voice or specific communication standards.

Understanding and navigating these limitations is key to effectively integrating AI into business writing processes. When used judiciously, it's a powerful tool, augmenting human expertise rather than replacing it and requiring a blend of AI efficiency with human insight for optimal results.

Mitigating the Limitations

Recognising the limitations of ChatGPT in business writing shouldn't act as a deterrent; rather, it should empower you to utilise this tool more adeptly. By crafting precise prompts and implementing a thorough review process, you can produce business content that meets technical standards and stands out for its quality and relevance. This involves understanding ChatGPT's capabilities, such as its ability to quickly generate drafts and provide diverse perspectives, while being mindful of its limitations like data recency and lack of industry-specific nuances.

The key lies in a symbiotic relationship with the AI. It's about acknowledging where ChatGPT falls short – such as in understanding the unique context of your business or the subtleties of your brand voice – and capitalising on its strengths, like speed and scalability. This approach enables you to craft the AI's output into something technically sound, engaging, thoughtful, and well-suited to your needs.

Think of ChatGPT as a versatile collaborator in your business writing endeavours. It's not a silver bullet for perfection, but it can significantly streamline your writing process when used strategically. This involves a balanced blend of AI-driven efficiency and your creativity and insight, ensuring the final content resonates with your intended audience and aligns with your business objectives.

In essence, the effective use of ChatGPT in business writing hinges on a close integration of AI capabilities with human oversight. This will ensure that the final output is both high-quality and contextually appropriate.

AI is Not a Mind Reader

I've used ChatGPT extensively to collaborate with me on this book. I've used many different chats to create the content, but it's crucial to understand that ChatGPT, like any AI, is not a mind reader.

It doesn't inherently grasp the intricate details of your business model or the specific requirements of your project. Providing explicit and detailed instructions is essential for obtaining relevant and tailored outputs. For instance, a vague command like "Write me a book about using ChatGPT to write a book" will not yield a document that fully aligns with objectives or writing style. ChatGPT requires context and clear direction to be effective.

The AI operates based on the information provided in prompts, as you've seen so far, making the quality and specificity of these prompts critical. Detailed inputs about your business's niche, target audience, goals, and any other relevant specifics can significantly enhance the relevance and usefulness of ChatGPT's outputs. This means formulating prompts that concisely yet comprehensively convey your needs.

Furthermore, understanding the limits of ChatGPT's knowledge base is important. As of November 2023, ChatGPT now has access to the Internet; it can generate images and will soon be able to accept voice commands and share chats containing images, probably by the time you are reading this. It's about leveraging AI as a tool that can assist with generating ideas, drafting content, and providing a base to build upon – but not as a substitute for human expertise and decision-making.

In summary, the effective use of ChatGPT in business contexts hinges on recognising that while AI is a powerful tool, it functions best when guided by detailed, well-structured prompts and supplemented by human expertise.

The Importance of the Right Prompt

I've previously defined prompts and how to use them to research what you want to learn about. The prompt is your way of directing ChatGPT. A vague or too-general prompt will likely result in a vague or too-general output. However, a well-thought-out prompt can lead to high-quality content tailored to your needs. But crafting that better prompt might require a few tries. Don't get discouraged; even seasoned writers need to edit and revise.

Authenticity and Originality

ChatGPT, despite its advanced design to produce unique responses, is not immune to generating content that may not be entirely original. It synthesises responses based on vast amounts of training data, sometimes leading to the creation of content that echoes or closely resembles existing sources. This aspect is particularly crucial in a business context, where the authenticity and originality of content are paramount, especially for material that is business-critical or intended for public dissemination.

To safeguard against unintended plagiarism and ensure the uniqueness of the content, it is advisable to run any AI-generated material through a plagiarism checker. This step is essential for verifying originality and maintaining the integrity and credibility of your business communications. It's about the balance between leveraging AI's efficiency and ensuring that the output aligns with the standards of originality and authenticity that your business upholds.

Additionally, Appendix A provides more comprehensive guidelines and best practices for effectively utilising plagiarism-checking tools with AI-generated content. This resource is designed to assist you in ensuring that the material you use is efficient and effective and maintains the highest standards of originality.

The Human Element

In the rapidly evolving landscape of AI, it's important to remember that no machine, as of now, can fully replicate the depth of human creativity, emotional intelligence, and the unique perspectives that come from individual experiences. While ChatGPT is an exceptional tool for drafting, brainstorming, and even generating initial ideas, it should be viewed as a complement to human capabilities rather than a substitute.

Think of ChatGPT as a versatile assistant offering a starting point, suggesting alternatives, or providing a framework for your thoughts. However, the true essence and value of any writing in a business context come from the human touch. This includes the ability to understand and connect with specific audiences, the application of emotional nuances, and the incorporation of unique insights and experiences that AI cannot replicate.

Your final output should be a blend of AI-generated drafts and your own insights. It involves a process of refinement where AI's efficiency in generating content meets the human skills of contextual understanding, emotional depth, and strategic thinking. This blend ensures that the final piece is not just technically sound but also resonates with your audience, reflecting the values and voice of your brand or business.

Moreover, integrating AI into your workflow should enhance, not diminish, the human elements in your work. It's about leveraging technology to augment creativity and productivity, ensuring the final product is innovative and deeply human-centric.

In essence, the optimal use of AI in business writing seamlessly combines the speed and breadth of AI with the irreplaceable depth and emotional intelligence of human expertise.

Conversing with a Digital Mentor

While ChatGPT is a machine, envisioning it as a knowledgeable mentor or subject matter expert can greatly enhance the quality and depth of your interactions. It's a shift from transactional queries to meaningful dialogue. By ascribing a persona to the AI, even temporarily, you're not just tapping into a database of facts—you're engaging in a conversation with a "specialist."

For instance, if you want to understand complex human behaviours, frame your question as if you're seeking advice from a seasoned psychologist. An example of a prompt might be something like:

> *Using the persona of a psychologist, could you explain the nuances of pathological lying and how I might recognise such behaviour in coaching sessions?*

Caution: Asking ChatGPT to answer using this persona doesn't make you the expert. You must carefully weigh the response before offering ChatGPT's response to anyone. It is, however, an excellent way to gather contextual information quickly.

This approach sets a context, guiding the AI to produce more tailored and insightful answers. It's like having a room full of experts, each ready to offer their perspective. Whether you need advice from a historian, a scientist, a linguist, or any other specialist, adjust your framing and dive into the conversation.

While ChatGPT is not human, treating it as one during interactions can make your dialogue richer, more nuanced, and surprisingly human-like.

Seeding the AI Idea: An Unplanned Journey

There's a fascinating allure to those "Aha" moments - the split second when an idea strikes, feeling so profound that it's as if a light bulb has physically been turned on. That moment came as I was deep in one of my sessions with ChatGPT. The depth, breadth, and dynamism of our interactions made me ponder:

What if the best way to write a book about ChatGPT was, paradoxically, to co-author it with ChatGPT?

But where to start? A structured, meticulously planned roadmap? That's typically the recommended route, but it felt counterintuitive. Instead, I trusted the process, and much like a sculptor facing a block of marble, I took a chisel and started... anywhere.

With each prompt, each query, the form began to emerge. There were hits and misses, of course. But the agility of the AI meant that with every iteration and refinement, the narrative became more compelling. Sometimes, it meant rewriting entire sections, while a subtle tweak made all the difference at other times.

This wasn't about perfection; it was about authenticity. It was about capturing the spontaneity of conversation, the rawness of an idea's birth, and the art of iterative crafting. The process was as much a discovery for me as I hope the content will be for you.

Where to Start?

When I first thought about writing a book, this topic wasn’t my first choice. I posed a question to ChatGPT, having already done some cursory research into my proposed topic. The more I researched, using a combination of ChatGPT, Bing, and Google, the less confident I was that my idea would be of sufficient interest to warrant the time and energy. Within the same chat, I posed another question.

> *“I’ve just had what I think is almost an epiphany! I’m determined to write about my experience with Generative AI and ChatGPT, and one way to do that is to collate and curate my 200+ chats into a small book. The content is relevant despite, and almost because of, the rapid acceleration in AI tools. The working title for my book will be* The Little Book of Chats: A Practical Guide to Generative AI and ChatGPT. *What do you think?”*

It's not my intention to include a full transcript of every chat because many of them are not pertinent; however, determining how to craft each prompt and understanding what I expect to get as a response is fundamental to the benefit you, too, can derive from the approach. An abridged snippet of ChatGPT’s response is shown below.

Wow, Peter, that sounds like a ripper of an idea! *The Little Book of Chats: A Practical Guide to Generative AI and ChatGPT* is a working title that immediately signals the book's focus and offers a fresh, up-to-date perspective on Generative AI. It gives the impression of being both approachable and informative, like a handy guide.

The next task was to write an introduction. Here's the prompt I gave ChatGPT.

Given that you now know quite a bit about my background and experience, I'd like your help crafting the introduction to the book. For example, fifty-five years in business, co-author of a university-level textbook, a university lecturer, a master's degree, an Applied Coaching certificate, and the boldness to make a career pivot at the age of seventy-two.

When I review a response from ChatGPT, I'm looking for several things:

1. Does the response read well
2. Does it sound like something I would say
3. Does it sound believable?

NOTE: I don't, and I strongly recommend that you don't just accept any response without testing the three facets above.

NOTE: Items 2 and 3 above are mutually exclusive. Just because you think it sounds like something you would say doesn't necessarily mean it's believable.

I didn't like ChatGPT's Introduction enough to use it. I didn't like ChatGPT's attempt at a preface enough to use it, either. There are some parts of both sections that I did use, but not verbatim.

The moral is that ChatGPT is a great tool for gathering content, using it for research and gaining differing perspectives, but deciding what to use is still very much a function of the human mind.

In Chapter 2, I delve into the impact I think AI will have. I'm not alone in this. Respected coaches are writing and blogging about how AI influences the coaching, mentoring, and education professions.

Before continuing, look at this article from the *Oxford Review – Using Artificial Intelligence to Enhance Coaching – A new study* (Using Artificial Intelligence to Enhance Coaching - A new study (oxford-review.com)).

Figure 3 - Image generated by ChatGPT-4, 27th November 2023

DALL-E Image Description

An abstract monochrome image depicting the impact of AI on the personal and professional lives of various personas, including a female, a male, a professor, and a business person. This image conveys a sense of disruption and challenging behavioural change, using interconnected lines, shapes, and textures to symbolise the transformative effect of AI.

CHAPTER 2 – AI's Impact On Everything

AI is not just a futuristic concept; it's a transformative force seamlessly integrated into our present, emerging seemingly overnight with profound impacts. Its rapid development and widespread adoption have revolutionised numerous industries, from healthcare to finance, and even how we approach everyday tasks. This swift evolution of AI has made it a crucial part of the current technological landscape, reshaping how we work, communicate, and make decisions.

This sudden advent of AI technology, especially in its more advanced forms like machine learning and deep learning, has implications far beyond mere technical advancements. It challenges our traditional understanding of tasks that can be automated or augmented by technology. The ripple effect of AI's integration is seen in everything from business operations and strategic planning to personal productivity and creative endeavours.

AI's presence and influence are undeniable, and its impact extends into societal, ethical, and economic realms. It has prompted important discussions about job automation, data privacy, and ethical AI use. As we stand at this juncture, the role of AI in shaping the future becomes a critical area of exploration, necessitating a thoughtful approach to leveraging its potential while navigating its challenges.

AI, in essence, has shifted from being a distant vision to a tangible, influential part of our daily lives and work, marking a pivotal moment in the technological era.

AI's Impact on Coaching

This topic is particularly interesting to me and, I suspect, to many other coaches, mentors, and educators. Within my coaching and mentoring practice. I actively encourage my clients to use ChatGPT, as discussed earlier.

Coaches possess a unique skill set in asking powerful questions to stimulate introspection and self-discovery in their clients. They are adept at designing questions that encourage deeper reflection and unlock new perspectives.

Like coaches skillfully design questions to guide clients toward greater self-awareness and growth, prompt engineering involves crafting prompts that facilitate meaningful exploration and discovery.

Coaches can use AI, like ChatGPT, as a complementary tool rather than a replacement. It can enhance coaching by providing additional resources and facilitating self-reflection. Coaches who embrace AI may see it as a way to augment their services, not diminish them.

Teaching clients to ask themselves powerful questions using AI may empower them. However, it's essential to remember that coaching isn't just about asking questions; it's about the coaching relationship, accountability, support, and the human element. Clients may still value coaches' guidance and expertise, even if they can ask powerful questions themselves. An AI isn't going to check in on you to see whether you've started to make the behavioural changes that your interaction with it suggested you should.

Coaches need to adapt to evolving client needs and preferences. Some clients may indeed use AI for self-reflection, but others may still seek the guidance and personalised support that coaches offer. Coaches who understand these changing dynamics can position themselves as valuable partners in the AI-enabled coaching landscape.

Coaches also need to consider the ethical implications of AI integration. They should guide clients on using AI responsibly and understanding its limitations. This ethical responsibility can be a valuable part of the coaching process.

While some coaches may be concerned about clients becoming more self-reliant through AI, it doesn't necessarily imply that coaches resist teaching clients to use such tools. The key lies in recognising that AI can enhance coaching and focusing on leveraging it responsibly to benefit both coaches and clients.

There are ethical considerations as well. While AI can provide information, it doesn't necessarily ensure accuracy, relevance, or ethical use. Human coaches can guide their clients in navigating their fields' ethical and practical aspects.

AI can enhance self-guided learning and knowledge acquisition, which aligns more with mentoring. However, mentors' human connection, context, and personalised guidance remain valuable. The evolving role of coaches may involve teaching clients how to use AI tools effectively while providing the critical human touch that complements AI-driven learning.

Two leading coach educators, Professor David Clutterbuck and Jonathan Passmore, have written about how AI already impacts coaching.

In his analysis of the rapidly evolving coaching landscape, Professor David Clutterbuck (Clutterbuck, 2023, How Fast is Coaching Changing?) highlights three primary forces driving change: technology, client behaviour, and the coaches themselves. He emphasises the significant impact of technological advancements like AI and VR in coaching, predicting that these tools will become more user-friendly and realistic. However, Clutterbuck notes the challenge of keeping pace with such technology, especially considering the varying levels of tech adoption across different age groups and economic constraints.

A pivotal aspect of this transformation is the evolving role of AI in coaching, particularly its potential to automate basic coaching methodologies like the GROW model. This raises questions about the future of human-led coaching and the unique value it offers, especially in aspects AI cannot replicate.

The demographic and attitude of coaches towards new technology mirror the general population. Clutterbuck points out that current competency frameworks in coach training, which emphasise process-following, might soon become obsolete. He argues for a radical revamping of coach education to include systemic thinking, complex adaptive systems approaches, and technological literacy, integrating these with traditional coaching skills.

Predicting the exact trajectory of these changes is challenging, as they might not progress uniformly. Clutterbuck advises coaches to stay informed and adaptable in their personal development plans to remain relevant and ahead in this dynamic environment.

Jonathan Passmore's article (Passmore, 2023, Coaching education: Wake up to the new digital and AI coaching revolution!) delineates the evolution of coach education over three decades (1990-2020), identifying three distinct phases. The first, 'pre-profession', was characterised by ad hoc and non-qualification-based training. The second phase, 'practice-based professionalisation', witnessed growth in small-scale coach providers utilising professional body competencies. The third phase, 'evidence-based professionalisation', marked the emergence of university-based coach education programs focused on evidence-based and research-informed training.

As we move into the mid-2020s, Passmore observes a shift from 'professionalisation' to 'productisation' in the coaching industry.

This new era is defined by the rise of large-scale, digitally enabled coaching providers employing thousands of home-working coaches. These providers aim to deliver coaching at scale to many coachees in large organisations, primarily through digital channels.

This industrial change necessitates a modernisation of coach education. Passmore emphasises the importance of coach educators recognising the transition towards managing industrial-scale delivery, integrating AI into the coaching process, and mastering technologies enabling global coaching operations. He proposes that coach education should offer two new career pathways: one for those managing coaching services and another for those working in digital coaching firms in roles like Customer Success and Coach Relations. Additionally, he suggests a revitalised coach training that equips coaches to operate in digital environments, mastering communication platforms, tools, and apps.

AI's Impact on Mentoring

Coaching often focuses on behavioural change and personal growth, while mentoring typically involves knowledge sharing and guidance. The role of human connection, support, and behaviour change is often more pronounced in coaching.

AI, including prompt engineering techniques, can empower individuals to engage in self-guided learning and problem-solving. When clients can access information and guidance independently, it can reduce their reliance on a mentor for knowledge acquisition.

In the mentoring context, AI can indeed change the dynamics. Mentors may find themselves more in the role of facilitating access to resources and guiding mentees on how to use AI tools effectively. This shift can lead to a more efficient transfer of knowledge.

However, the human connection in mentoring remains valuable. Mentors provide context, insights, and personalised guidance that AI alone cannot replicate. Additionally, mentors offer mentorship beyond just knowledge transfer, such as networking opportunities and career advice.

There are ethical considerations as well. While AI can provide information, it doesn't necessarily ensure accuracy, relevance, or ethical use. Human mentors can guide mentees in navigating the ethical and practical aspects of their fields.

AI can enhance self-guided learning and knowledge acquisition, which aligns more with mentoring. However, mentors' human connection, context, and personalised guidance remain valuable. The evolving role of mentors may involve teaching mentees how to use AI tools effectively while providing the critical human touch that complements AI-driven learning.

In her blog post, Mentoring in the AI World (Schnieders, 2023), she addresses how integrating AI into mentoring programs brings forth transformative benefits, reshaping the traditional mentor-mentee dynamic. Key aspects of this integration include:

1. **Enhanced Pairing Mechanisms**: Leveraging AI's data analysis capabilities, mentor-mentee matches are more precisely aligned based on shared objectives and interests. This precision fosters more impactful and successful mentoring relationships.
2. **Customization of Developmental Journeys**: AI's ability to evaluate individual characteristics and learning preferences allows for the creation of bespoke developmental pathways for mentees. This personalised approach ensures that each mentee receives guidance that resonates with their unique aspirations and learning styles.
3. **Expansion and Efficiency**: AI's automation and analytical skills enable mentoring programs to expand their reach

while maintaining quality. By handling administrative complexities, AI facilitates the management of numerous mentoring pairs, ensuring consistent quality and insight-driven program enhancements.

4. **Insightful Program Analysis**: AI's analytical prowess extends to evaluating the effectiveness of mentoring programs. It provides comparative benchmarks and identifies best practices, offering a deeper understanding of what drives successful mentor-mentee interactions.
5. **Skill Enhancement Through Technology**: AI also plays a pivotal role in the skill development of participants. Interactive AI tools focus on refining critical skills like problem-solving and communication, providing targeted exercises and feedback, thus enriching the mentoring experience.

In summary, AI's role in mentoring is multifaceted, from improving matching accuracy and personalising learning experiences to enhancing program scalability, offering data-driven insights, and aiding in skill development. This technological integration marks a significant leap in how mentoring programs are designed and executed.

AI's Impact on Education

The integration of AI in education has the potential to significantly enhance the learning process and reshape the teaching profession in several ways.

AI can adapt to individual student needs by assessing their strengths, weaknesses, and learning styles. It can provide tailored content, practice exercises, and feedback, allowing students to progress independently. This shift from a one-size-fits-all approach to personalised learning can improve student engagement and outcomes.

AI-powered tools can make education more accessible to a broader range of students. It can accommodate different learning abilities, including those with disabilities, by providing customised resources and support. This inclusivity is a key aspect of AI's impact on education.

Routine administrative tasks, such as grading assignments and managing schedules, can be automated with AI, freeing educators to focus more on teaching and mentoring. This can lead to more efficient use of educators' time and resources.

AI can analyse vast amounts of data related to student performance, identifying trends and areas for improvement. Educators can use these insights to make data-informed decisions about curriculum design and teaching strategies, ultimately enhancing the quality of education.

AI-powered educational platforms can provide learning opportunities around the clock. Students can access resources, ask questions, and receive assistance anytime, which is particularly beneficial for adult learners, remote students, or those with busy schedules.

AI can assist educators in providing more targeted guidance and mentorship to students. It can identify struggling students early and suggest interventions. This enables educators to focus their mentorship efforts where they are most needed.

AI can serve as a teaching assistant, helping with routine tasks, answering common student queries, and providing instant feedback. This allows educators to scale their impact and support more students effectively.

Educators can use AI to access the latest research, teaching methodologies, and best practices in their field. AI can recommend relevant resources, enabling educators to stay updated and improve their teaching skills.

AI-powered collaboration tools can facilitate global connections among educators and students. This fosters cross-cultural learning experiences, breaking down geographical barriers in education.

With AI handling some administrative tasks and content delivery, educators can shift their focus to mentoring, coaching, and fostering critical thinking and problem-solving skills in students. This evolution of the educator's role is an opportunity for professional growth.

Across these domains, AI empowers professionals to evolve their roles. Coaches become even more effective by leveraging AI's capabilities in prompt engineering. Mentors shift toward facilitating AI-driven learning and providing context. Educators focus on mentoring, critical thinking, and global collaboration.

AI's positive impact hinges on ethical and responsible use. Properly implemented, AI complements human expertise, enhances learning, and fosters personal and professional growth. It empowers individuals to explore, learn, and adapt in an ever-evolving world.

AI's Impact on Innovation

The advent of AI challenges the traditional human-centric approach to innovation. AI's capacity to process and analyse vast datasets surpasses human capabilities, enabling it to uncover patterns and correlations that might elude human analysis. This advanced analytical ability positions AI as a powerful tool in innovation, particularly in identifying opportunities for new products and services.

AI can reveal insights from data that are complex, multidimensional, and often beyond the scope of human comprehension. For instance, in industries like pharmaceuticals, AI can expedite the discovery of new drugs by analysing countless molecular combinations, a task that would be immeasurable for human researchers. In consumer products, AI can analyse consumer behaviours and preferences at a granular level, leading to the development of highly tailored products.

Furthermore, AI's predictive analytics can forecast emerging market trends, allowing businesses to stay ahead of the curve. It also plays a crucial role in optimising product design and manufacturing processes, enhancing efficiency and sustainability.

However, this shift raises important considerations about the role of human creativity and intuition in the innovation process. While AI can identify patterns and generate data-driven insights, the human ability to understand context, ethics, and emotional nuances remains vital. Balancing AI's analytical strengths with human creativity and ethical judgment will be key to harnessing the full potential of AI-driven innovation.

Summary

AI empowers professionals to evolve their roles. Coaches become even more effective by leveraging AI's capabilities in prompt engineering. Mentors shift toward facilitating AI-driven learning and providing context. Educators focus on mentoring, critical thinking, and global collaboration.

AI's positive impact hinges on ethical and responsible use. Properly implemented, AI complements human expertise, enhances learning, and fosters personal and professional growth. It empowers individuals to explore, learn, and adapt in an ever-evolving world.

Figure 4 - Image generated by ChatGPT-4, 27th November 2023

DALL-E Image Description

A pencil drawing depicting a female author sitting at a computer drafting a book. Beside her is a blank pad with an indecipherable list of topic headings on the screen, represented by abstract markings. The drawing captures her in deep concentration, focused on her work, highlighting the creative writing environment.

CHAPTER 3 – Using ChatGPT for Business Writing

In this chapter, I explore the multifaceted realm of AI applications in business writing, a domain where the blend of technology and language opens new opportunities for efficiency and creativity. AI, particularly models like ChatGPT, has made significant inroads into various business writing facets, offering substantial benefits and notable challenges.

One of the most common applications of AI in business writing is in drafting emails. ChatGPT can efficiently generate professional, well-structured emails, saving time and enhancing communication clarity. Report writing is another area where AI proves invaluable. It can assimilate and summarise data, providing insightful reports that would otherwise demand considerable human effort and time.

Beyond these conventional uses, AI is great at creating marketing content, such as product descriptions, blog posts, and social media updates. Its ability to quickly generate creative, engaging content makes it a powerful tool for digital marketing strategies. Similarly, in public relations, AI can assist in drafting press releases and communication briefs, ensuring consistent messaging and tone.

AI's capabilities extend to more technical aspects of business writing, such as drafting business plans and proposals. It can help structure these documents, suggest content based on industry trends, and even aid in competitive analysis. For businesses involved in research and development, AI can assist in preparing research summaries and white papers, synthesising complex information into readable formats.

AI finds its use in legal writing in more specialised domains, helping draft basic legal documents and contracts. While it doesn't replace legal expertise, it provides a starting point that legal professionals can refine. Another area is policy writing, where AI can help articulate company policies, ensuring they are clear and comprehensive.

However, this increasing use of AI in business writing comes with a set of challenges and considerations. The limitations of AI in understanding context, maintaining originality, and aligning with a specific brand voice are crucial factors. Additionally, ethical concerns about transparency in AI-generated content and the need for human oversight are paramount.

This chapter explores how AI, particularly ChatGPT, transforms business writing, examining its benefits, limitations, and ways to harness its potential while emphasising the human aspect effectively.

Business Case Proposal Example

I created a prompt asking ChatGPT to generate a business case proposal to see how this works in practice. This may not be a perfect example, but it took about two minutes to think about and write the prompt and a further five minutes to transfer the prompt and response into this chapter.

> *I'd like to try writing a high-level business case proposal, not the business case itself. This should be about modelling a set of business processes, customer journeys, requirements (system, functional, and non-functional), role descriptions, a RACI matrix, and the relationship between processes and/or process steps and functional requirements. The approach is to combine these artefacts into a digital tender to be published for vendor responses. Because this will go to tender, there is no need to focus on financial factors until we complete the more detailed model. Can you do this in 500-600 words, please?*

High-Level Business Case Proposal: Digital Tender for Business Process and Customer Journey Modelling

Introduction

In today's rapidly evolving business landscape, the need for streamlined and efficient business processes has never been more critical. This proposal outlines a project to model a set of business processes and customer journeys, with the aim of creating a comprehensive digital tender for vendor responses. Our focus is to amalgamate various business elements - system requirements, functional and non-functional requirements, role descriptions, a RACI matrix, and the interrelation between processes and functional requirements - into a unified digital format.

Objectives

To Model Business Processes and Customer Journeys: Develop detailed models that encapsulate our current business processes and customer journey maps, highlighting areas for optimisation and efficiency.

Define System and Functional Requirements:

Clearly articulate the system, functional, and non-functional requirements that underpin our business operations, ensuring they align with our strategic objectives.

Role Clarification and RACI Matrix Development:

Establish clear role descriptions and develop a RACI (Responsible, Accountable, Consulted, Informed) matrix to ensure clarity in responsibility and decision-making processes.

Align Processes with Functional Requirements:

Analyse the relationship between our business processes/steps and functional requirements to ensure cohesion and efficiency.

Preparation of a Comprehensive Digital Tender: Compile the above elements into a digital tender document, ready for distribution to potential vendors for their responses.

Approach

Our approach involves a step-by-step methodology, starting with the mapping and modelling of existing business processes and customer journeys. This will involve a thorough examination of current workflows, identifying bottlenecks and potential areas for improvement.

Subsequently, we will define system requirements and both functional and non-functional requirements. This will ensure that the technological solutions proposed by vendors align seamlessly with our business needs.

In parallel, we will delineate clear role descriptions and develop a comprehensive RACI matrix. This matrix will serve as a guideline for responsibilities and accountability within the project, ensuring efficient decision-making and execution.

A critical aspect of our approach is the alignment of business processes with functional requirements. This will involve a detailed analysis to ensure that each process step directly contributes to meeting these requirements, thereby ensuring operational efficiency and effectiveness.

Finally, all these elements will be combined into a digital tender document. This document will serve as a call for proposals from potential vendors, who will provide their insights and solutions based on the detailed information provided.

Expected Outcomes

The expected outcome is a detailed, clear, and comprehensive digital tender that effectively communicates our requirements to potential vendors. This will enable us to assess the vendors' capabilities and solutions accurately, ensuring we select a partner that can meet our specific business needs.

Conclusion

This high-level business case proposal sets the groundwork for a transformative project that aligns our business processes with the technological capabilities available in the market. By approaching this through a detailed digital tender, we position ourselves to find the best possible solutions, driving efficiency, and innovation in our business operations.

Next Steps

Following the approval of this proposal, the next steps will involve assembling a project team, initiating the process modelling phase, and developing a timeline for the subsequent stages of the project.

Assuming the Business Case Proposal outlined above is successful, the next step is gathering your content and preparing the Business Case itself. The following prompt exemplifies how you might choose to do this. Let's see how ChatGPT responded to this request.

Writing the Introduction to a Business Case

Could you guide me on what to include in a prompt for creating an introduction to a business case? I understand the introduction is crucial and should set the right tone for the rest of the document. What key elements should I include to ensure it's comprehensive and effective? I think it should outline the

> *business problem or opportunity, provide context or background information, and briefly mention the objectives or goals of the case. Also, how should I structure this information in the prompt to get a well-constructed introduction from ChatGPT?*

To give this the right context, ChatGPT responded by giving me a few headline topics and then a revised version of the prompt. The purpose of the prompt revision was so the output would be specific to the business case proposal rather than the generic response you see here.

1. **State the Business Problem or Opportunity**: Begin your prompt by clearly defining the specific problem or opportunity your business case will address. This sets the focus for the introduction.
2. **Provide Context and Background Information**: Include essential background details that give a snapshot of the situation. This might involve industry context, market trends, or previous efforts related to the problem or opportunity.
3. **Mention Objectives or Goals**: Briefly outline what the business case aims to achieve. This sets expectations for the proposed solutions or strategies.
4. **Structure of the Prompt**: Structure your prompt logically. Start with the business problem or opportunity, followed by background context, and then state the objectives.

Here's an example of how to structure your prompt for ChatGPT:

"I'm drafting an introduction for a business case and need guidance. The case addresses [briefly state the business problem or opportunity]. It's set in [provide a brief background or context, like industry or market conditions]. The primary objectives are [state the goals or what you aim to achieve]. Could you create an introduction based on this information?"

This format ensures ChatGPT receives all the necessary information to generate a relevant and targeted introduction for your business case.

The Value Proposition

As I've just shown, this is at least a comprehensive and reasonable first draft based on the prompt I gave ChatGPT. You could ask for another version, for certain sections or passages to be reworded or rewritten, or for more or less detail by changing the word count (this came out at 536 words). As I mentioned in Chapter 1, don't expect perfection.

Here are four things that you should take into account when considering the usefulness of this approach.

1. **Enhancing Efficiency and Productivity**: One of the primary advantages of using ChatGPT in business writing is the significant boost in efficiency and productivity. ChatGPT can quickly generate drafts, provide ideas, or even create entire documents, saving valuable time. Discussing how this speed can be leveraged to meet tight deadlines, handle a high volume of communication, or free up human resources for more complex tasks would be beneficial.

2. **Customization and Personalization**: ChatGPT's ability to tailor content based on specific prompts means businesses can create highly customised and personalised written content. This includes crafting emails catering to individual client needs, creating targeted marketing material, or generating

reports aligning with specific internal guidelines. Exploring the balance between AI's efficiency and the need for human touch in personalising content can provide a nuanced view of its application in business writing.

3. **Maintaining Quality and Accuracy**: While ChatGPT can produce content quickly, there is a need to ensure the quality and accuracy of the output. Addressing how businesses can implement review processes and guidelines to maintain high standards in AI-generated content is crucial. This includes fact-checking, ensuring brand consistency, and adapting the AI's output to suit the company's tone and style.

4. **Ethical Considerations and Transparency**: In the context of business writing, it's important to discuss the ethical implications of using AI-generated content. This includes transparency about the use of AI in creating content, especially in public-facing documents, and understanding the ethical boundaries regarding content ownership and authenticity.

These key areas provide a comprehensive overview of the benefits and challenges of using ChatGPT in business writing, offering a well-rounded perspective on effectively integrating AI into business communication strategies.

Writing a Business Book

I referred earlier to using ChatGPT to write this book. Using ChatGPT or similar AI tools for writing a business book is a creative and innovative approach. However, there are important considerations to keep in mind to ensure the process is effective and the final product is of high quality. Here are five key caveats for using ChatGPT in writing a business book:

1. **Verify the Accuracy of Information:** While ChatGPT is trained on a vast array of texts and can provide generally accurate information, it's not infallible. Especially in a business book, where factual accuracy is crucial, independently verify any facts, statistics, or specific business advice the AI provides. Remember, the training data only goes up to a certain point (April 2023 for ChatGPT-4), so it may not be up-to-date with the latest trends or data.
2. **Maintain Your Unique Voice:** AI can mimic various writing styles, but ensuring your book retains your unique voice and perspective is essential. This is especially important in a business book, where personal insights, experiences, and a distinctive voice can greatly enhance the reader's engagement and the book's credibility. Use ChatGPT to aid your writing, not to replace your personal style and tone.
3. **Be Aware of Potential Bias:** AI models can inadvertently reproduce biases in their training data. When using ChatGPT for topics involving opinions, societal norms, or cultural aspects, be mindful of potential biases in the responses. Always cross-check and ensure the content aligns with a balanced and fair perspective, especially when dealing with sensitive topics.
4. **Customize and Personalize Content:** While ChatGPT can generate content based on the input provided, it's crucial to customise and personalise this content to align with your book's specific goals, audience, and approach. AI-generated content should be seen as a starting point or a draft to be further refined and tailored to your book's needs.
5. **Use AI for Creativity, Not Just Information:** ChatGPT can be a great tool for brainstorming, generating ideas, or overcoming writer's block. It can suggest different angles for a topic, provide examples, or even help structure the book. However,

always add your analysis, insights, and creative touch to these suggestions to make your book insightful and original.

AI tools like ChatGPT are powerful aids in the writing process, but the most effective and impactful books will always have a strong human element – your expertise, experiences, and creative vision are what will truly resonate.

Was This Book Written in a Single Chat?

The simple answer is 'No'. When I was creating prompts to gather information about specific topics, I split them into multiple chats. I did this to make finding or extending content easier when proofreading.

For example, I have a separate chat, which I use to generate the images you see throughout the book. Writing a prompt to generate an image is more complex than a prompt to solicit information. ChatGPT doesn't generate images; that's done through DALL-E. I've acknowledged that in naming each figure. Many of those images have been regenerated, and I also refined the prompts to obtain an image more closely aligned with the picture I had in my head for that part of the book.

What's the best approach?

I don't pretend to be a novelist, but I was curious to know what the difference would be if I were writing a novel compared to this book. I asked ChatGPT the following question:

Can you explain the difference in approach if I had been using ChatGPT to write a novel versus a book about using ChatGPT to write a book about ChatGPT? I'm particularly interested in the structure you would have outlined for each book and how you would have suggested I start the process.

Writing a Novel Using ChatGPT

Approach:

1. **Conceptualization:** Start with a clear idea or theme for your novel. This could be a specific genre, a plot outline, or a set of characters you wish to explore.
2. **Structure Planning:** Novels typically follow a structured narrative arc—introduction, rising action, climax, falling action, and resolution. Plan your novel's structure, perhaps chapter by chapter, outlining key events or character developments.
3. **Character Development:** Flesh out your characters using ChatGPT. You can brainstorm backstories, characteristics, and motivations for each key player in your narrative.
4. **Dialogue and Scenes:** Use ChatGPT to help write dialogue, describe settings or draft scenes. AI can be particularly useful in generating creative descriptions or natural-sounding dialogue.
5. **Feedback and Refinement:** Regularly revise and seek feedback on your writing. ChatGPT can assist in editing for grammar, style, and coherence, but human input is crucial for a novel's creative and emotional aspects.

Starting the Process:

Begin with a simple prompt to ChatGPT describing your novel's basic idea or a specific aspect you want to develop, like a character sketch or a scene description.

Writing a Book about Using ChatGPT to Write a Book

Approach:

1. **Defining the Scope:** Clearly outline what aspects of using ChatGPT you want to cover. This might include how to generate ideas, develop content, edit drafts, and the limitations of AI in creative writing.
2. **Structural Organization:** This book would likely follow a more instructional or guidebook format. Organise it into chapters or sections based on different uses of ChatGPT (e.g., brainstorming, drafting, editing).
3. **Practical Examples:** Include real examples of interactions with ChatGPT. This could involve showing how specific prompts lead to different types of responses or how AI can contribute to various stages of writing.
4. **Research and Analysis:** Incorporate research on AI's role in creative writing, discussing its potential and limitations. This might involve looking at current literature on AI in creative processes.
5. **User Guide Approach:** Adopt a tone and structure similar to a how-to guide, providing readers with actionable advice on effectively utilising ChatGPT in writing.

Starting the Process:

Start by drafting an outline of the key topics and chapters. Your initial interaction with ChatGPT could involve asking it to generate a basic outline or to provide insights on the best practices for using AI in writing.

General Differences in Approach:

- **Novel Writing:** The focus is on storytelling, character development, and narrative arcs. The use of ChatGPT is more creative and imaginative, helping to craft engaging narratives and compelling characters.
- **Instructional Book Writing:** The emphasis is on informative, practical content. The structure is more linear and segmented, with each section aiming to educate the reader on a specific aspect of using ChatGPT. The use of ChatGPT here is more analytical and demonstrative.

In both cases, it's essential to remember that ChatGPT is a tool to assist in the writing process. For a novel, it aids in the creative aspects, while for an instructional book, it serves more as a subject of analysis and demonstration. Regarding creativity, analysis, and personal experience, the human touch remains central to both types of writing projects.

Consider ChatGPT as a Virtual Digital Assistant

As shown throughout the book, ChatGPT can act in many capacities.

1. **Assistance with Tasks:** Just like a digital assistant, ChatGPT can help with various tasks, from generating ideas and writing drafts to providing suggestions for revisions. It can

take on the heavy lifting for certain aspects of writing, much like an assistant would in a traditional work setting.

2. **Interactivity:** ChatGPT offers an interactive experience. You can have a back-and-forth dialogue, ask questions, refine your queries based on previous responses, and build upon the information it provides, similar to how you would work with a human assistant.
3. **Time-Saving:** ChatGPT can save time in the writing process, much like a good assistant helps streamline work. It can quickly generate content, offer suggestions, and even help with research, accelerating the writing process.
4. **Supplementing, Not Replacing, Human Input:** Just as a digital assistant isn't a replacement for human decision-making and creativity, ChatGPT supplements the writing process. It's a tool that aids in creativity and productivity but doesn't replace the essential human elements of writing, such as personal voice, nuanced storytelling, and critical judgment.
5. **Limitations and Guidance:** Like any digital assistant, ChatGPT has its limitations and works best when given clear instructions and guidance. Understanding these limitations and learning to guide the AI in your interactions effectively will yield the best results.

Productivity, efficiency, accuracy, and integrity. What more can you ask?

Performing a Final Edit

In addition to having three external reviewers of the material in this book, I wanted to check for duplicated content and the flow of information from chapter to chapter. I created a prompt to do that, section by section.

Why not upload the whole book and do it in one go?

I've previously referred to limits imposed by ChatGPT regarding the amount of content you can upload into a single prompt. Here's the prompt.

I want to give you chapters of my book to review. I'm specifically looking for duplicated content, not literally, but where I'm saying something very similar. If you find duplication, please identify it and suggest an amendment, without changing the tone, intent or word count.

Without showing the response to the prompt for each section, here's a summary of the response to the first section I uploaded.

Thank you for sharing the Preface of your book. After reviewing the content, I did not find any significant duplication within this section. Each paragraph introduces a distinct aspect of your experience and perspective, effectively setting the stage for what the book will cover.

This exercise took less than fifteen minutes, including writing, copying, and pasting the prompts and responses into this section. In a book of 140+ pages, that's a significant productivity boost. I'm pleased to say that ChatGPT found no duplications worth noting and commented that the flow of material from Chapter to Chapter was logical and consistent. You will be the best judge of that.

What's the Benefit of Doing This?

There are several reasons why you should consider incorporating AI-assisted tools like ChatGPT in your book writing or business documentation processes:

1. **Demonstrating Efficiency**: Utilise AI tools to validate information and check for duplication efficiently. This approach is particularly beneficial in business writing, where time is a valuable resource, and accuracy is crucial.

2. **Applying AI in Editing and Proofreading**: Leverage AI for editing and proofreading. It showcases a practical application of AI in ensuring content originality and integrity, significantly streamlining the editing process.

3. **Highlighting the Complementary Role of AI**: Understand that AI is a tool to complement your efforts, not replace them. It can efficiently handle tasks like scanning large volumes of text, allowing you more time for creative or analytical tasks.

4. **Real-Time Interaction and Feedback**: Engage in real-time interaction with AI for immediate feedback and suggestions. This feature is invaluable in business environments where quick decision-making is often required.

5. **Educational Insights:** Gain educational insights into the practical use of AI in real-world scenarios. It's one thing to understand AI's capabilities theoretically; experiencing it in action is quite another.

6. **Reflecting on the AI-Assisted Writing Process**: Reflect on how AI-assisted writing can evolve and be utilised at various

stages of content creation, from drafting to final proofreading.

7. **Encouraging AI Tool Adoption**: By incorporating AI tools into your writing process, you not only enhance your work but also encourage others, especially those in business and executive roles, to consider similar integrations.

8. **Embracing Innovation**: Show your commitment to innovation by continuously exploring and integrating the latest AI tools in your writing process. This approach keeps your content dynamic and aligns with the evolving landscape of AI technology.

Incorporating AI into your writing process offers a blend of efficiency, accuracy, and innovation. It allows you to focus on the creative aspects of your work while AI handles the more mundane tasks, ensuring that your final product is both high-quality and timely.

Next, we embark on a practical journey, guiding you through the art of creating your own prompts, a skill pivotal for leveraging AI tools like ChatGPT to their fullest potential.

Figure 5 - Image generated by ChatGPT-4, 27th November 2023

DALL-E Image Description

An abstract monochrome image depicting a person engaging with ChatGPT on a computer. This design represents the process of creating prompts, using abstract forms to symbolise the interaction between the individual and the AI technology.

CHAPTER 4 – Creating Prompts for Yourself

There are many topics you might choose to explore using ChatGPT. I can't begin to imagine what they might be, so in this chapter, I'll outline topics I'm interested in and demonstrate how I crafted the prompts. If they don't resonate with you, take a moment to think about what's important to you and create your own. I'll provide examples for both Business Writing and Personal Research.

NOTE: ChatGPT cannot access files on your internal network. It cannot connect to a database to retrieve corporate information. If this is important to your organisation's use of Generative AI, OpenAI publishes an API I mentioned briefly in the Introduction. Third-party providers may be able to provide solutions using the API to access internal information, but that's outside the scope of this book.

Business Writing Examples

Chapter 4 marks a transition into a more interactive and personalised exploration of AI's capabilities in business writing, focusing on the art of creating prompts. Creating tailored prompts is not just an exercise in directing AI like ChatGPT; it's a journey into understanding your unique business needs and how to articulate them effectively. This chapter aims to empower readers with the skills to craft prompts that resonate with their specific goals and challenges, thereby maximising the potential of AI in their professional sphere.

The primary benefit of crafting personalised prompts lies in aligning AI outputs with specific business objectives. When a prompt is tailored to reflect the nuances of a business scenario, the AI's response is more likely to be relevant, actionable, and valuable. For instance, prompts carefully designed to include industry-specific language, desired outcomes, and contextual details can guide AI to produce outputs that are accurate and resonant with a business's unique brand voice and strategic direction.

Crafting precise prompts becomes even more crucial in more specialised or esoteric areas. For example, in niche markets like technology startups or boutique consulting firms, the use of industry-specific jargon and context in prompts can lead ChatGPT to generate insights or content that is significantly more targeted and applicable. This specificity can extend to tasks like drafting patent applications, where the technical language and detailed descriptions in prompts can steer the AI towards generating more precise and relevant content.

Furthermore, well-crafted prompts can guide AI to simulate various business scenarios in strategic planning and decision-making, offering insights tailored to a business's unique challenges and opportunities. This can extend to areas like risk assessment, market analysis, and employee training, where specific prompts can help generate focused content, scenarios, or training modules.

Throughout this chapter, we will explore exercises and examples that demonstrate how to create effective prompts and illustrate the transformative impact of these prompts in harnessing AI's capabilities for personalised business writing solutions. The goal is to provide readers with the tools and insights to make AI an integral and highly effective part of their business communication arsenal.

Create a Blog Post

Give ChatGPT an outline of what you want to blog about. Give it a suggested title and ask whether that is consistent with your topic. Three key elements should be outlined:

1. **Possible Title**: This is crucial as it sets the theme and tone, drawing readers in. It should be engaging, succinct, and reflective of the main content, often incorporating SEO-friendly keywords.
2. **Information About the Post**: This encompasses the central topics, objectives, key messages, and specific arguments or perspectives. Detailing the purpose (whether to inform, persuade, or entertain), addressing key questions, and highlighting unique insights are essential. Including relevant research or data points is also beneficial.
3. **Tone, Cadence, and Style**:
    - **Tone**: The overall attitude or character of the writing, be it formal, conversational, or humorous, tailored to the audience and message.
    - **Cadence**: The rhythm and pace influence readers' engagement with the content.
    - **Style**: Encompasses language choice, sentence structure, and approach, aligning with the target audience's preferences and understanding level.

I asked ChatGPT to create a prompt to demonstrate what you might write. Here's its response.

Suggested Prompt

Title: 'Harnessing AI in Modern Business Practices'. I'm planning a blog post about integrating AI technologies in business operations, focusing on efficiency improvements, ethical considerations, and future trends. The post should examine how AI reshapes business strategies and the workforce and discuss balancing technological advancement with ethical responsibility. I aim for a professional yet engaging tone with a clear and concise style accessible to a general business audience. Does this title accurately reflect the content and intent of my post?

Project Reports and Executive Summaries

Creating project reports and executive summaries is crucial to business communication, providing stakeholders with a concise overview of project outcomes. ChatGPT can assist in distilling complex project data into clear, concise reports. Users can provide information about project objectives, methodologies, key results, challenges encountered, and lessons learned. ChatGPT can then structure this information into an organised format, highlighting significant achievements and areas for improvement. This is especially useful for project managers and team leaders who regularly report on project progress and must communicate effectively with stakeholders who may not have detailed technical knowledge of the project.

To assist effectively with the preparation of Project Reports and Executive Summaries, a user would need to provide detailed and specific information, including:

1. **Project Overview**: A brief project description, including its purpose, scope, and objectives. This sets the context for the report.
2. **Key Activities and Milestones**: Information about significant activities undertaken during the project and milestones achieved. This helps in outlining the project's progress.
3. **Data and Results**: Quantitative and qualitative data gathered during the project. This could include performance metrics, survey results, or other relevant findings.

4. **Challenges and Solutions**: Details of any challenges encountered during the project and the strategies or solutions implemented to address them.
5. **Outcomes and Impact**: A summary of the project's outcomes and impact. This might include achieving goals, benefits to the organisation or stakeholders, and any learnings or insights gained.
6. **Future Recommendations**: Suggestions or recommendations for future actions based on the project's results and learnings.
7. **Specific Requirements for the Summary**: For the executive summary, specify any particular focus areas or points that must be highlighted, keeping in mind that it should be a concise and comprehensive overview of the entire project.
8. **Tone and Style Preferences**: Indicate if there's a specific tone or style to adhere to, aligning with organisational standards or the report's intended audience.

Providing this detailed information will enable ChatGPT to assist in crafting a well-structured and informative project report and executive summary tailored to the specifics of your project and organisational needs.

Speech Writing and Public Speaking

ChatGPT can be a valuable asset in speech writing, offering a starting point for crafting speeches for various occasions. Users can specify the event, the audience, the key messages they wish to convey, and the desired tone, whether formal, motivational, or humorous. ChatGPT can then generate a draft with an engaging introduction, a coherent body, and a memorable conclusion. This is particularly helpful for individuals who have clear ideas but may struggle with the structure and flow of a speech. The draft provided by ChatGPT can serve as a base, which the user can then personalise, adding anecdotes, personal insights, and specific references to make the speech more relatable and impactful.

For example, creating a prompt to prepare a speech for a Toastmasters or Rostrum event involves specifying key details about the speech, its objectives, and the audience. Here's an example of how you might structure such a prompt:

> *Please help me prepare a speech for an upcoming Toastmasters/Rostrum event. The event's theme is [insert theme], and my speech should be about [insert specific topic]. I aim to [state your objectives – inform, persuade, entertain, etc.]. The audience will primarily be [describe the audience – their general age, interests, Toastmasters/Rostrum experience level, etc.]. I want the speech to be engaging and [mention any specific tone or style – humorous, inspirational, informative, etc.]. The speech should be approximately [state the duration] minutes long. Could you provide a structure for the speech and suggest key points I should cover?*

This prompt gives ChatGPT clear instructions on the speech's topic, objectives, audience characteristics, tone, and length, allowing for a more tailored and relevant response. Remember, the more specific and detailed your prompt, the more targeted ChatGPT's assistance will be.

Meeting Summaries and Minutes

ChatGPT can be invaluable in transcribing and summarising key points from business meetings. Users can input their notes or recorded audio from a meeting, and ChatGPT can generate a clear, concise summary or formal minutes. This includes identifying key decisions, action items, and discussion highlights, ensuring that important information is captured efficiently and can be easily referenced by team members post-meeting.

Can ChatGPT really accept recorded audio? At the time of writing, the answer is an emphatic 'No'; however, this can be accomplished in several ways. Since the official recognition of the COVID-19 pandemic in March 2020, technology has rapidly evolved and improved in scalability to provide communication capabilities for people who could no longer meet face-to-face in their usual office environments.

Zoom, Microsoft Teams, Google Meet, and Otter.ai are examples of meeting software that can record and transcribe audio. This is not an exhaustive list but is used to demonstrate how the process works.

After a meeting, your tool of choice can transcribe the audio to text, and this is the start of turning the recording into your meeting minutes and action plans.

NOTE: Many factors affect the accuracy of any transcription service, for example, background noise, different dialects, participants' speaking volume, etc.

NOTE: There is an ethical consideration that can't be ignored. This has nothing to do with ChatGPT per se and more to do with those feeding the transcribed audio into ChatGPT. For example, a coaching session recording audio using Otter.ai was to be submitted for assessment by a mentor coach in conjunction with the video recording. Strict instructions were given to those participating in the process that spelling mistakes could be corrected, but grammatical errors could not, i.e., the recording had to represent a verbatim copy of the coaching session.

How could you instruct ChatGPT to address any spelling mistakes generated through the transcription without making grammatical improvements?

That prompt might look something like this.

> *Please review the following text, which is a transcription of a meeting held on [date] with the following attendees [attendee list]. You may correct any spelling or typographical errors caused by the transcription, but you must not change the output's content, tone, or grammatical style.*

There is still a potential ethical issue: trusting the person using ChatGPT to transcribe the recording without materially changing the final content.

This requires a process to store the original audio file and the completed transcription from ChatGPT, which has to be done outside ChatGPT. A link to the edited chat could be saved with the original audio file in an agreed location.

Personal Research Examples

But what if you want to use ChatGPT for things of a more personal nature?

ChatGPT offers a unique, private platform for individuals to explore various personal interests, questions, and even emotional inquiries. Its capacity to provide information, generate ideas, and simulate reflective conversations makes it a versatile tool for personal exploration and learning. For instance, while ChatGPT is not a substitute for professional counselling, it can serve as a preliminary space for individuals to articulate and explore their emotions and thoughts. By framing questions about emotional well-being or personal challenges, users can gain insights or perspectives that might help in understanding their feelings better.

When it comes to educational pursuits, ChatGPT can be an invaluable resource. Users looking to expand their knowledge or explore further study options can use AI to gather information on various subjects, academic programs, and learning opportunities. It can provide summaries, explanations, and even reading suggestions on topics ranging from literature to sports psychology. This makes it a useful starting point for deeper academic exploration or satisfying personal curiosity.

However, users should be aware of the limitations of using ChatGPT for personal research. While it offers a breadth of information, its responses are based on pre-existing data and may not always reflect the most current research or nuanced viewpoints. Additionally, it's important to consider the value of human judgment and professional expertise for sensitive or deeply personal topics, which AI cannot replicate.

In navigating these uses, crafting precise and thoughtful prompts becomes key. Users should approach ChatGPT with clear questions or topics to ensure relevant and focused responses. Whether seeking information on a new hobby, understanding a complex psychological concept, or exploring personal growth avenues, framing questions can significantly influence the quality and usefulness of the AI's responses.

Understanding and Addressing Imposter Syndrome

Most of us will experience feelings of inadequacy at least once, if not several times, throughout our lives. How would you ask ChatGPT to help you understand and address this? You could start with a prompt like this:

> *I'm experiencing a sense of apprehension and inadequacy about my current role. Can you give me some advice about how I conduct a reality check so I know what I need to work on?*

ChatGPT might respond with a list of things for you to consider, for example:

1. Address your current situation
2. Identify feelings of inadequacy
3. Analyse your skills and knowledge
4. Evaluate past performance
5. Consider external factors
6. Seek feedback
7. Set realistic goals
8. Plan for development
9. Reflect on achievements and growth

Each of these items is the trigger for another question for ChatGPT. For item 2 above, you might create a prompt like this:

> *In your role as my mentor, how do I determine the causes of my feelings of inadequacy?*

For item 3, you might create a prompt like this:

> *I've heard some products and services can help me determine my strengths and natural talents. Can you give me the names of a few? Are any of these free?*

Based on the response you receive from ChatGPT, you might ask additional questions. Keeping these in the same chat so ChatGPT has the underlying context for each new prompt is a good idea.

The Psychology of Musical Chairs

This may seem a little left-field, but I was watching a children's show with my granddaughters. The episode was about playing musical chairs, and I wondered what the psychology behind the game is, i.e., what's it teaching children? I used that as a prompt.

This is a summary of the response. The game of musical chairs has intriguing psychological aspects. They are:

1. Competition and Winning
2. Quick Decision-Making and Adaptability
3. Observation and Strategy
4. Handling Disappointment and Resilience
5. Physical Activity and Coordination
6. Social Interaction and Cooperation
7. Rules and Fair Play

These are only the subheadings, with additional information for each one. You can ask ChatGPT to expand on any of these topics, and for those of you who've had children, there's a lot to learn from a seemingly innocuous game.

Using ChatGPT as a Personal Mentor

In Chapter 2, I talked about the impact AI will have, or is already having, on Mentoring. I conducted an experiment where I asked ChatGPT how I could encourage my mentoring clients to use ChatGPT for their personal research.

I created a prompt and asked ChatGPT whether that would be an appropriate way for someone to start interacting with it.

Please act as my personal mentor. I'll give you some background information, and I'd like you to pose some questions that will help me open up about what I want to learn.

ChatGPT can ask a reflective question to start the process, for example, "Tell me what you're currently doing professionally".

Note that there's some role reversal here. ChatGPT is now asking the questions, and the user is responding. Let's assume that the client comes back with something like this:

> *I currently work as a senior business analyst with one of the major banks here in Australia. I've been here about four years, but I've started looking around for roles that will extend my capabilities.*

ChatGPT's response might be:

- "That's a good chunk of time you've spent at your current role. What specifically makes you feel that you're ready to extend your capabilities?"
- "Have you identified the types of roles or projects that interest you? If so, what are they?"
- "What skills or qualifications do you think you'd need to acquire to transition successfully?"

Each question forms the basis of another response from the client, and at some point, the roles might again be reversed, and the client will start asking questions rather than responding to those asked by ChatGPT.

Understanding and Utilising Your Strengths

One of the first things I did when I started coaching was to learn more about my strengths. I undertook the VIA Character Survey, the CliftonStrengths Profile, the DiSC 2.0 Profile, and the Enneagram Profile. They each have a unique perspective and provide insights into who you really are.

I prefer the CliftonStrengths Profile, and I encourage my clients to do this as part of their coaching or mentoring experience with me. I created a prompt to understand how ChatGPT would respond before giving my clients a version of that prompt they could use to examine their own strengths.

> *I want to explore how to utilise my strengths, specifically those from my CliftonStrengths profile. I'll give you my top five strengths and would like you to suggest how I best utilise these in my personal and professional life.*

This became an extensive interaction with ChatGPT, resulting in 11,322 words or approximately 40 A4 pages. It's relevant because I went back and revisited the chat many times, asking new questions that led to different answers. I don't know how detailed my coaching or mentoring clients' chats were, but their feedback was that they discovered alternative opportunities to those they'd thought about before using ChatGPT.

Explore Your Areas of Interest

Even in a business setting, your interests may be personally motivated. For example, you've recently been promoted to lead a significant project, and you soon notice that your team is not functioning as well as it might.

You might construct a prompt something like this.

As a newly promoted manager of a team of six business analysts working on a significant project, I'm facing challenges with stakeholder clients who are resistant to change. This resistance is impacting my team, leaving them feeling unsupported. Could you provide strategies for building my negotiation skills to better support my team and facilitate more effective interactions with these stakeholders? I'm particularly interested in techniques for managing resistance to change and ensuring my team feels supported.

You might recognise the symptoms of self-doubt and want to know how you can address that. You might construct a prompt something like this:

I've recently been promoted to manage a team of business analysts, but I find myself doubting my abilities to lead them successfully. Despite my boss's confidence, I often feel like I might not be cut out for this role and fear failing. This seems like imposter syndrome. Can you provide insights on how to overcome these feelings? I'm looking for strategies to boost my self-confidence and practical tips for developing my leadership skills to lead my team effectively and overcome these doubts.

Here's another more personal prompt.

> *I excel in problem-solving and understanding complex issues but struggle with memorising facts, a barrier to courses and certifications requiring closed-book exams. This is limiting my career advancement opportunities. Can you provide advice or strategies to improve my memory of facts, especially in the context of an exam? Additionally, are there techniques or study methods that cater to someone with my skill set, who is more adept at hands-on problem-solving than rote memorisation?*

Here are five personal topics where ChatGPT can provide valuable advice, strategies, or techniques:

1. **Self-Confidence Building**: ChatGPT can offer strategies and exercises to boost self-confidence, including positive self-talk, setting achievable goals, and techniques to build a positive self-image.
2. **Managing Stress and Anxiety**: The AI can suggest techniques for stress management and reducing anxiety, such as mindfulness practices, breathing exercises, and time management tips.
3. **Improving Communication Skills**: ChatGPT can provide tips on enhancing interpersonal communication skills, including active listening, effective speaking, and non-verbal communication cues.
4. **Overcoming Fear of Public Speaking**: The AI can offer advice on public speaking, including how to prepare and manage nervousness and techniques for engaging an audience effectively.

5. **Developing Healthy Habits**: ChatGPT can suggest ways to establish and maintain healthy habits, such as regular exercise, a balanced diet, and effective sleep routines, contributing to overall well-being.

These topics are well-suited for AI assistance as they involve general guidance and techniques that have been widely discussed and validated.

Your challenge is to craft the right prompt to give you information you can use. If any of these topics resonate with you, ask ChatGPT for guidance or advice.

ChatGPT can be a powerful tool for personal research and exploration, provided you approach it with clear intentions and an understanding of its capabilities and limitations.

Figure 6 - Image generated by ChatGPT-4, 27th November 2023

DALL-E Image Description

An abstract monochrome image depicting the transformation from apprehension about AI into a vast array of business opportunities. The design includes an abstract figure and elements that suggest growth and increased performance, conveying a narrative of progression and optimism in the business context.

Chapter 5 – Transforming Apprehension into Opportunity

I think we all understand the meaning of 'The Elephant in the Room', particularly regarding fear about AI. Fear is a multi-layered psychological phenomenon. It often stems from the anxiety of the unknown, the perceived threat to existing power structures, and a deeply ingrained scepticism towards change.

AI is far from achieving a nuanced understanding of human emotions, cultural subtleties, and complex social contracts that guide our behaviour, but not perhaps as far as we think. The human touch is irreplaceable in professions like coaching, mentoring, psychology, counselling, and human resources, where personal chemistry and understanding can be as critical as facts and techniques.

AI has become a powerful tool across various industries; the coaching field is no exception. As a business and executive coach, I acknowledge my bias because this is where I work and where I see opportunities to utilise AI for the benefit of my clients. While there may be concerns within the coaching community about AI replacing human coaches, it is crucial to address these misconceptions and shed light on the immense potential for AI to enhance and complement the coaching profession. Let's explore the positive benefits of applying AI to coaching and demonstrate how they far outweigh the perceived negatives.

A Philosophical Perspective

A recent study called Trust in Artificial Intelligence: A Global Study (Gillespie et al., 2023) examined the extent to which people trust AI systems.

The relationship between trust and fear in the context of AI is indeed a fascinating subject. Trust in AI is fundamentally about people's confidence in the technology's reliability, safety, and fairness. Fear often stems from uncertainty, a lack of understanding, or concerns about potential negative outcomes, such as job displacement, loss of privacy, or uncontrolled AI behaviours. The developmental acceleration of AI makes it difficult to keep up with, and trust is easily lost.

Education plays a crucial role in transitioning from fear to trust. By increasing understanding, education demystifies AI technology. It can clarify how AI works, its limitations, and its potential benefits, which helps to alleviate unfounded fears. Moreover, education about AI can foster critical thinking, enabling individuals to differentiate between AI's real challenges and sensationalised myths.

Here are a few thoughts on how this transition can be facilitated through education:

1. **Literacy and Fluency**: Basic AI literacy programs can inform the general public about what AI is and is not capable of, reducing fear through understanding. More in-depth fluency for professionals can help them to interact competently with AI in their fields.
2. **Ethical and Societal Implications**: Teaching about the ethical use of AI and its societal impacts can foster discussions on ensuring equitable AI solutions designed with societal welfare in mind.
3. **Hands-on Experience**: Practical experience with AI, such as creating prompts about topics that interest them, can help people understand AI's functionality and limitations.
4. **Policy and Regulation Education**: Understanding how AI is regulated can reassure people that safeguards against potential misuse or errors exist.

5. **Fostering a Culture of Continuous Learning**: Given AI's rapid evolution, instilling a mindset of lifelong learning can help people stay informed and adaptive to technological advancements.

By educating the public and stakeholders on these fronts, the transition from fear to trust can be smoother, fostering an environment where AI is viewed as a tool for enhancing human capabilities rather than a nebulous threat. This education should be multifaceted, including the workings of AI and its implications for society, to fully address the "elephant in the room."

Here are a few other things to consider.

1. **The Unknown:** The psychology of fear is often rooted in the ambiguity surrounding the impact of AI. Coaches, executives, and academics may find predicting how AI will change the game's rules challenging. This uncertainty fuels their fear and makes them wary of AI adoption.
2. **Existential Threat:** AI poses a potential threat to job security and the uniqueness of human skills. This adds a layer of existential dread, where individuals fear their role or expertise becoming obsolete.
3. **Systemic Belief Challenge:** Traditional frameworks represent established norms. People inherently fear that AI could disprove or make these frameworks less relevant. Hence, they are resistant to even discussing it.
4. **Certification vs Experiential Competency:** Many people equate professional validity with certification. This ingrained mindset makes them wary of AI, which promotes data-driven, experiential competency over traditional certification. The resistance is stronger among those who have invested time and resources in gaining these credentials.

One of the key principles of coaching is helping the client establish a growth mindset. Coaches, mentors, and educators must be open to the same growth mindset themselves if they are going to apply AI to their coaching and mentoring toolkit.

Addressing the fear of AI among this group requires a carefully crafted narrative. It's important to show them that AI is not a replacement but a supplement that can enhance their human expertise. Case studies that successfully integrate AI in traditional settings might be persuasive. Address their anxieties by highlighting how AI can offer new avenues for demonstrating and acquiring competency far beyond the limitations of standard certifications.

A recent update to ChatGPT supports referencing (see below), which you'll recall I specified earlier in my discussion about custom instructions. **NB: *Source generated as part of ChatGPT's response*.**

Source: *Epley, N., & Caruso, E. (2009). Perspective taking: Misstepping into others' shoes. In K. D. Markman, W. M. P. Klein, & J. A. Suhr (Eds.), Handbook of imagination and mental simulation (p. 295–309). Psychology Press.*

The irony is palpable when professionals who are supposed to facilitate change resist it themselves. Let's break this down a bit:

1. **Cognitive Dissonance:** One explanation for this paradox could be cognitive dissonance. Coaches and psychologists understand change at an intellectual level, but they struggle when the change implicates their own practices and beliefs. They might experience discomfort when their professional identity or values are threatened.
2. **Loss Aversion:** Humans prefer avoiding loss to acquiring equivalent gains. The potential loss of traditional methods,

reputation, or even clientele could make coaches and psychologists hesitant to adapt to new approaches like AI.

3. **Expertise Bias:** Professionals build a sense of expertise and authority in their fields over time. Introducing a game-changing element like AI can make them feel like novices all over again, which is an uncomfortable position.
4. **Control Aspect:** Facilitating change in others can sometimes give the illusion of control over change itself. When the change is external and challenges their modus operandi, that sense of control is threatened, making it psychologically uncomfortable.
5. **Social Validation:** There's also a herd mentality at play. Until a critical mass within their profession begins to accept and adapt to new methods, many coaches and psychologists will be hesitant to be the early adopters, fearing professional ostracization.
6. **Defensive Pessimism:** Psychologists often understand defensive pessimism — the strategy where people envision the worst-case scenario to manage their anxiety. This may be a coping mechanism where they view AI as the worst-case disruptor to their career and prefer to maintain the status quo.

So, the irony essentially boils down to human nature and inherent cognitive biases. Recognising this can be the first step in self-awareness for these professionals, allowing them to confront their fears and embrace change, much like they advocate for their clients. **NB:** ***Source generated as part of ChatGPT's response***.

Source: *Festinger, L. (1957). A Theory of Cognitive Dissonance. Stanford University Press; Kahneman, D., & Tversky, A. (1979). Prospect Theory: An Analysis of Decision under Risk. Econometrica, 47(2), 263-292.*

Continuing in this vein, much like psychologists can understand depression yet still be susceptible to it, coaches and psychologists can be well-versed in theories of change and yet struggle with it themselves.

1. **Psychologists and Depression:** Knowing the signs and treatments for depression doesn't make psychologists immune to experiencing it. The same applies to their attitudes towards change. Just because they have the tools to facilitate it doesn't mean they find it easy to apply those tools to themselves. They're not immune to the human condition, after all.
2. **Coaches and Prescriptive Approaches:** Coaches trained and accredited in specific frameworks can become prescriptive in their approach. This "toolkit" of skills and methods becomes their safety net, and the idea of integrating AI or other innovative techniques threatens this established comfort zone.
3. **Risk of Dogma:** In both cases, a rigid attachment to existing frameworks can lead to a kind of professional dogma. When this happens, the individual's effectiveness can be compromised as they become less adaptive and more resistant to evolving methodologies, including AI.
4. **Cognitive Lock-in:** Both psychologists and coaches can suffer from "cognitive lock-in," where their training makes them so committed to a certain way of thinking that they fail to adapt, even when presented with more effective options.

This duality, where professionals are resistant to the very change they are trained to facilitate in others, illustrates the complex relationship we all have with change. It also offers an opportunity for these professionals to 'walk the talk' by embracing the same principles of adaptability and openness that they advocate for in their practices. **NB:** ***Source generated as part of ChatGPT's response.***

> **Source:** *Bandura, A. (1999). Social Cognitive Theory: An Agentic Perspective. Asian Journal of Social Psychology, 2(1), 21-41; Aronson, E., Wilson, T. D., & Akert, R. M. (2013). Social Psychology (8th ed.). Pearson Education Inc.*

Whether you're a coach, a mentor, a psychologist, a counsellor, or anyone involved in changing the behaviour of your clients to help them realise improvements and move forward, here are some final suggestions.

1. **Mirror Test:** Coaches need to do a self-assessment, asking questions that mirror the ones they usually ask their clients. Questions like, "How adaptable are you?" or "When was the last time you stepped out of your comfort zone?" can be powerful catalysts for self-reflection.
2. **Status Quo Bias:** The inherent irony of advocating for change while resisting it. Examine whether you're clinging to outdated practices due to a status quo bias.
3. **The AI Reality:** AI is not a futuristic concept but a present reality. Your clients are already engaging with AI in various facets of life and will increasingly expect you to be knowledgeable in this area.
4. **A Challenge for Change:** Set a challenge to integrate at least one AI tool or methodology into your practice within a set timeframe. This could be as simple as using machine learning algorithms to analyse client feedback better or as complex as incorporating AI-driven virtual coaching as a supplementary resource.

5. **Accountability:** Just as you hold your clients accountable, who holds you accountable for your own growth and adaptation?
6. **Inversion Strategy:** Try thinking about the inversion strategy. What is the worst-case scenario if you don't adapt? Often, confronting the negative consequences can be a powerful motivator for change.

Examine your own attitudes towards change - both behavioural and technological. After all, the best coaches are also lifelong learners. **NB: *Source generated as part of ChatGPT's response.***

Source: *Kahneman, D. (2011). Thinking, Fast and Slow. Farrar, Straus and Giroux; Munger, C. (1995). Poor Charlie's Almanack: The Wit and Wisdom of Charles T. Munger. Donning Company Publishers.*

You may have heard of or read the book *Feel the Fear and Do It Anyway (Jeffers, 1987)*. This could be an excellent framework to explore how people, particularly professionals, view AI. Jeffers' main tenets could be quite illuminating when applied to the context of technological fear:

1. **Understanding Fear Isn't the Problem:** Jeffers argues that fear will always exist but shouldn't paralyse you. Similarly, many professionals fear AI but don't necessarily understand it. Your book can stress the importance of engaging with this fear rather than avoiding it.
2. **Taking Responsibility:** Jeffers urges readers to take responsibility for their reactions. Coaches and psychologists can't control AI's progress, but they can control how they respond to it. You can challenge your readers to take

responsibility for their professional growth in an AI-driven world.

3. **Making No-Lose Decisions:** Jeffers describes every decision as a "no-lose" situation because even wrong decisions provide valuable lessons. Adopting AI into one's coaching or professional practice could be framed similarly to a learning experience, irrespective of immediate outcomes.
4. **The Whole Life Grid:** Jeffers suggests dividing life into segments like work, leisure, and relationships to focus on improvements in each area. Professionals could adopt a similar multi-faceted approach to AI integration, exploring its potential in different aspects of their practice.
5. **Action Cures Fear:** Jeffers is big on the idea that taking small steps can gradually help overcome fear. Your book could offer actionable ways for professionals to gradually incorporate AI into their work, mitigating or reducing their apprehension through practice.
6. **Choosing Love and Trust:** Lastly, Jeffers discusses choosing love over fear, which can be translated into choosing to trust that technological advancements like AI can be beneficial, not just detrimental.

Embrace the opportunity to transform each of Jeffers' principles into actionable challenges or tasks related to AI. This book combines Jeffers' renowned psychological insights with cutting-edge AI expertise and coaching wisdom, offering you a powerful framework to confront and harness any apprehension you might feel about AI in your professional life. By actively engaging with these principles, you're taking a significant step towards turning potential fears into a source of strength and innovation.

I wanted to explore this further, so I asked ChatGPT another question.

What Happens When AI is Misused?

For those who want to subvert the positive benefits and focus on using AI for nefarious purposes, it probably won't matter much what form of governance is put in place. My thinking is that many of the products and technologies that appear or evolve over the next few years will be based on APIs, and those APIs cannot be as easily governed, i.e., someone who uses the OpenAI AI API (like ChatGPT-4) can add other functionality to it that is not the responsibility of Open AI. They can basically do what they want. A colleague of mine has built an app based on the API, but he stores data in a secure database and queries it using the ChatGPT-4 API. His intentions are not malicious, but even so, he's probably too small for any authoritarian governance to be applied. For years, we've seen smartphones take and edit pictures, and now we have things like magic erasers. My point is that some of the unsubstantiated or uneducated fears can't be mitigated by governance and can only reasonably be addressed by education. What's your view on this?

The response is lengthy, so I'll break it into more readable pieces.

The tension between innovation and governance, particularly with technologies as potent as AI, is a significant challenge. The point about APIs is spot on. Even if the core technology, like ChatGPT-4, adheres to ethical standards and guidelines, the potential for misuse can rise exponentially once it's integrated into third-party apps or systems via APIs. This is particularly true for smaller developers or entities that might fly under the regulatory radar.

We've seen reliably reported examples of deep fakery in which well-recognised people appear to be doing, saying, or suggesting something they did not do.

The issue is essentially the 'Pandora's Box' of technology; once it's open, it's hard to control who uses it and for what purpose. This is where the idea of education comes into play. Just like "Feel the Fear and Do It Anyway" advocates for taking personal responsibility for one's choices, a cultural shift is needed to emphasise ethical considerations in tech development and usage.

Coaches, psychologists, and other professionals can be critical in driving this cultural shift.

The rapid advances in AI technology have allowed ordinary people to do things without constraint with these capabilities. What will governance look like in this environment?

Governance alone can't mitigate all risks or fears. Laws and regulations are often slow to adapt, and they struggle to keep pace with the speed of technological innovation. Education has to be the first line of defence. This would entail instructing people on how to use the technology responsibly and educating them about its limitations, thereby dispelling unfounded fears.

Who Will Lose Their Jobs?

One of the most talked-about fears is the loss of jobs. That will almost certainly happen, so it's important to go back in time and see whether we can learn from the history of other technological changes. I asked ChatGPT this question.

What professions are most likely to be impacted by job losses caused by all forms of AI?

The impact of AI on job markets is a complex and evolving issue. Certain professions are more susceptible to automation and AI integration, potentially leading to job losses or significant changes in job roles. Some of the most likely affected professions include:

Manufacturing and Production Workers: Automation and robotics have significantly impacted these sectors. AI enhances the capabilities of these technologies, potentially reducing the need for human labour in routine and manual tasks.

Transportation and Warehousing: With the advancement of autonomous vehicles and drones, jobs in transportation (like trucking and taxi services) and warehousing may be significantly affected.

Retail: AI-driven technologies, including self-checkout kiosks and online retail algorithms, can reduce the need for cashiers and salespeople in brick-and-mortar stores.

Customer Service: AI chatbots and automated response systems are increasingly handling customer service tasks, potentially reducing the need for human operators.

Data Entry and Analysis: Jobs involving repetitive data entry and basic analysis will likely be automated through AI and machine learning algorithms.

Banking and Finance: Roles focused on routine tasks, like data processing and basic analysis in banking, can be automated. AI can also impact areas like financial advising with robo-advisors.

Agriculture: Automated machinery and AI-driven monitoring systems can reduce the need for manual labour in various agricultural processes.

Journalism and Writing: AI tools capable of generating news articles, reports, and even creative writing can impact jobs in journalism and content creation.

Healthcare Administration: Administrative roles in healthcare that involve scheduling, patient data management, and billing might see automation through AI systems.

Legal Services: Basic legal services like document review and case research can be automated using AI, impacting paralegals and junior legal staff.

It's important to note that while AI can replace certain tasks, it also creates new opportunities and job roles, especially for those who adapt and acquire skills to work alongside AI technologies. The future job market will likely value skills AI cannot easily replicate, such as creativity, complex problem-solving, emotional intelligence, and strategic thinking.

While the AI revolution may have awoken while we were sleeping, there are many examples throughout history of where change has necessitated a shift in our thinking.

Where Does This Leave Us?

The question of whether AI will lead us to a utopian or dystopian future is a complex and nuanced one. In its myriad forms, AI holds immense potential to revolutionise every aspect of our lives, offering solutions to long-standing problems and opening up new avenues for innovation and efficiency. However, it also poses significant challenges and risks, from ethical dilemmas to potential misuse.

The key to navigating this dual potential of AI lies in balancing its integration into society. This involves a collaborative approach where technological advancement is matched with ethical consideration, human oversight, and societal input. We must foster a symbiotic relationship with AI, harnessing its capabilities to enhance human endeavours while being vigilant about safeguarding our values and welfare.

In creating *The Little Book of Chats*, a concise guide of about 120 pages, I've endeavoured to bridge this gap. The content, a blend of my thoughts and ChatGPT's input – roughly 40/60 – mirrors the larger narrative of AI's role in our future. It demonstrates how human creativity and AI's efficiency can complement each other, producing work that neither could achieve alone. Reflecting on the question posed at the beginning of the book about the authorship ratio, it's clear that this collaboration embodies the essence of a balanced AI-human future.

This book, therefore, serves as a microcosm of the broader AI journey. It's a testament to the potential of AI and human collaboration and a reminder of the importance of steering this technology towards outcomes that benefit society as a whole, striving for a future that is neither purely utopian nor dystopian but grounded, balanced, and enriched by AI.

Summary

Generative AI is just one branch of the larger AI capability that invokes fear, uncertainty, and doubt in the minds of many. My own experience suggests that the ability to research, validate, and publish material through Generative AI is going to increase research productivity because even if the response isn't 100% accurate, you'll find that out much sooner and:

The older you get, the less time you have to learn new things!

Why have I placed such a heavy emphasis on coaching, mentoring, psychology, and counselling? AI is not dramatically different to people using "Dr Google" to diagnose their ailments or ask questions they don't want to ask real professionals. Apart from the odd disclaimer from doctors about the foolishness of doing so, people will search for a disease matching their symptoms. Equally, people can use Generative AI to get an answer that suits their purposes, not necessarily a balanced or valid one.

Denying the existence or impact of AI or failing to understand how it will affect you and your family, colleagues, and clients is naïve at best and potentially unprofessional.

Like a coach will say to their client, "Decisions about your future are up to you".

Similarly, what you choose to do with AI is up to you.

Figure 7 - Image generated by ChatGPT-4, 27th November 2023

DALL-E Image Description

A pencil drawing depicting a male and a female discussing the advanced features of ChatGPT-4 using a whiteboard. The drawing captures them in a collaborative pose, with abstract markings on the whiteboard to represent their discussion, free of any recognisable text. The sketch style aims to convey the essence of a brainstorming session, with detailed attention to their expressions and postures.

CHAPTER 6 – Advanced Features

Chapter 6 delves into the advanced features of AI and ChatGPT, particularly focusing on the capabilities that extend beyond basic text generation. This chapter aims to unveil the more sophisticated aspects of AI interactions, such as using Markdown for formatting, generating tables, and manipulating output for specific purposes. These advanced features enhance AI's usability in various contexts and open doors to more complex and creative applications.

Markdown, a lightweight markup language, is one feature that significantly elevates the utility of ChatGPT. It allows users to format text in a simple and readable way, making the content more engaging and organised. From creating bullet-point lists to embedding links and formatting headers, Markdown empowers users to structure their AI-generated content in a visually appealing and functionally effective way. This is particularly beneficial in contexts like report writing, content creation for web pages, and even in crafting detailed research summaries.

Another advanced feature is the AI's ability to generate and manipulate tables. This can be incredibly useful for data representation, summarising research findings, or organising complex information in a user-friendly format. In more esoteric areas, such as data analysis or project management, the ability to create structured tables can aid in presenting data-driven insights or project timelines clearly and concisely.

Moreover, ChatGPT's advanced features extend to its ability to tailor content according to specific tones, styles, or objectives. For instance, it can generate more persuasive content for marketing purposes or adopt a more empathetic tone for customer service interactions. This versatility makes it a powerful tool in diverse fields, from business communications to creative writing.

This chapter will explore these advanced features in-depth, providing insights into how they can be effectively utilised for various purposes. We will also look at potential esoteric applications of these features, like their use in academic research for synthesising complex theories, technology fields for generating technical documentation, or even artistic endeavours like poetry or scriptwriting.

The goal is to equip you with the knowledge and skills to leverage these advanced capabilities of AI, enhancing the scope and quality of your interactions with ChatGPT.

How Creative Are You?

What do you do with all the chats you create? Are they just for you, or do you want to use that information for another purpose, such as writing a book or a blog or contributing to a forum?

Some options change the way your responses are formatted. Even if you're not technical, you'll know what Hypertext Markup Language (HTML) is. It's the language used to render content on websites.

ChatGPT supports a capability called Markdown. This is a set of text formatting conventions used to denote specific styling or structural elements. You might choose to use some of these styles depending on how you want your output formatted.

Here's a link (Basic Syntax | Markdown Guide) to the complete list of markdown options; however, ChatGPT doesn't support all these. The following list highlights the more common markdown styles generally available in ChatGPT.

Note: You might still find it easier to do this in whatever software you use to format the output.

Here's a list of common formatting features you might find useful:

1. **Bold Text**: To make text bold, use two asterisks before and after the text. For example, **bold** renders as **bold**.
2. *Italic Text*: To italicise text, use a single asterisk or underscore before and after the text. For example, *italic* or _italic_ render as *italic*.
3. **Bullet Points**: Bullet points are created by starting a line with a hyphen (-) or an asterisk (*), followed by a space. For example:
    - Item 1
    - Item 2
4. **Numbered Lists**: Numbered lists are created by starting a line with a number, followed by a period and a space. For example:

1. First item
2. Second item

5. **Inline Code**: Using backticks before and after the text to indicate inline code or a command. For example, **`code`** renders as **code**.

6. **Block Quotes**: Start a line with a greater-than sign (>), followed by a space to create a block quote. For example:

> This is a block quote.

7. **Links**: To add a hyperlink, the format is **[link text](URL)**. For example, **[OpenAI](https://openai.com)** renders it as OpenAI.

8. **Images**: While ChatGPT doesn't embed images in its responses, markdown syntax for images is similar to links**`![alt text](image URL)`**.

9. **Headers**: Headers are created by placing several hash characters (#) before the header text, corresponding to the header level. For example, **# Header 1**, **## Header 2**.

10. **Strikethrough**: Using two tildes before and after the text to strike through the text. For example, **~~strikethrough~~** renders as ~~strikethrough~~.

11. **Square Brackets for Clarifications**: Square brackets often add clarifying information or editorial notes within a quote. For example: "He said he was [happy] to see me."

These formatting features can enhance the readability and structure of text-based communication, making it more effective, especially in a business context where clarity and presentation are important.

How Creative Do You Need To Be?

The question of creativity in using AI tools like ChatGPT is relevant, particularly if your aim extends beyond simple private or personal research. The need for creativity in crafting prompts is minimal for basic informational queries. However, this guide includes a discussion on creative prompt crafting because it's essential not to presume that everyone will intuitively know how to leverage these tools to their fullest potential.

Creativity in this context doesn't necessarily mean artistic creativity; it's about being thoughtful and strategic in communicating with AI. For more complex tasks, such as generating business content, creative writing, or developing nuanced research queries, framing your questions can significantly impact the quality and relevance of the AI's responses. Being creative here means considering what you ask and how you ask it – structuring your prompts to guide the AI towards the response you need.

Moreover, understanding the advanced features of AI, like using Markdown for formatting or tailoring responses for specific audiences, can enhance the effectiveness of your interactions. While these skills may not be necessary for straightforward research tasks, they become crucial when aiming to produce structured, audience-specific, or stylistically unique content.

While the basic use of AI for personal research might not require much creativity, exploring its more advanced capabilities to achieve specific goals or produce more refined outputs certainly does. This guide aims to provide you with the knowledge and tools to do just that, regardless of your starting point.

Code Generation (Programming)

ChatGPT's code generation capabilities represent a significant leap in the field of AI, offering substantial assistance in software development. This feature allows ChatGPT to write, analyse, and debug code, making it a valuable tool for novice and seasoned developers.

One of ChatGPT's code generation's key strengths is its language support versatility. It can work with various programming languages, including popular ones like Python, JavaScript, Java, C++, and HTML/CSS for web development. This makes it adaptable to various programming tasks, from scripting to full-scale application development.

Beyond just writing code, ChatGPT can also help debug existing code by identifying errors and suggesting corrections. This feature is particularly helpful for learning new programming languages or concepts, as it offers an interactive way to understand and fix coding mistakes.

Moreover, ChatGPT can assist in creating code documentation, explain complex programming concepts in simpler terms, and even guide users through developing algorithms or logic for specific tasks. However, it's important to note that while ChatGPT is a powerful tool, its code-generation capabilities should supplement human expertise. The AI-generated code often requires review and refinement by a human programmer to ensure accuracy and efficiency.

In summary, ChatGPT's code generation feature is a testament to AI's advanced capabilities in software development, offering support across various programming languages and tasks.

New Features in ChatGPT-4

At the beginning of the book, I explained that for most of the prompts you want to create, ChatGPT-3.5 would be sufficient. On the pages before the beginning of each new chapter and each appendix, you'll see what's referred to as a pencil drawing.

This feature is only available in ChatGPT-4, but I thought it would be useful for you to see how it works. Creating a prompt to generate an image is more complex than creating a plain text prompt.

My experience generating these images gave me a new appreciation of the complexity levels between text and image prompts. The best way to explain this is to ask ChatGPT.

Can you explain in 150-200 words why creating a prompt to generate an image is more complex and nuanced than creating a text prompt?

Creating a prompt to generate an image with AI, such as using OpenAI's DALL-E, is typically more complex and nuanced than creating a text prompt. This is due to the inherent differences between visual and textual information processing. In image generation with DALL-E, prompts must encompass multi-dimensional elements like colours, shapes, textures, and spatial relationships. For instance, specifying 'a sunset over a mountain' requires careful consideration of various visual factors to achieve the desired outcome. This contrasts with text prompts, which are linear and more direct. Additionally, visual interpretation varies widely among individuals, adding layers of subjectivity and requiring precise crafting of prompts. Furthermore, the technology behind image generation, like the Generative Adversarial Networks used in DALL-E, processes and generates complex visual data, demanding intricate prompts for effective guidance. Thus, the complexity of visual data, subjective interpretation, and advanced technology like DALL-E make image generation prompts more nuanced and challenging than text-based prompts.

I took ChatGPT's explanation of what the image contained and placed it below the caption. This roughly equates to the prompt, but many hits and misses exist. You may already have seen comments on the internet about images of people containing more fingers or limbs than are anatomically possible, so you need to be even more explicit in crafting a prompt to generate an image.

Even when you think you are being clear, the image may not represent what you asked for.

Using the following prompt, I asked ChatGPT-4 to generate an image.

I want a pencil drawing of Coach Peter with a golf club and a cup of hot chocolate.

Here's what was generated.

Figure 8 - Image generated by ChatGPT-4, 28th November 2023

The second attempt was even worse.

Figure 9 - Image generated by ChatGPT-4, 28th November 2023

I decided not to pursue this any further. However, I started using Bing to ask how I should structure prompts when I wanted to generate images. One advantage Bing has is that it typically generates four images based on the criteria outlined in your prompt, whereas ChatGPT-4 only gives you one.

ChatGPT-4 does not generate the image; that's done by DALL-E, despite my captions saying they were generated by ChatGPT-4. This is because you can also generate images directly using DALL-E outside ChatGPT-4; the results may differ.

It comes back to how precise you can be with the directions you provide in the prompt. I chose all my images to be monochrome because the book isn't printed in colour, but that's a specific attribute of the prompt. Otherwise, you're likely to get a colour image.

I also specified that I wanted pencil drawings, whereas I requested abstract images earlier in the book.

In Appendix D, I share several chats so you can see how I've created and augmented them over time. Unfortunately, at the time of writing, ChatGPT-4 doesn't support sharing chats containing images, so I'm unable to share the chats used to create images for the book. I'm sure that's on the drawing board for a future release.

OpenAI recently announced they have formally started the development of ChatGPT-5, so many of the challenges I've experienced with ChatGPT-3.5 and ChatGPT-4 will hopefully be resolved in the not-too-distant future.

How Long Did It Take to Write the Book?

In the beginning, I said it took twelve days to write the first draft. After feedback from my reviewers, much of the content was removed, rewritten, or moved to other chapters. Adding the images generated by ChatGPT took another few days while I learned how to create better prompts, and I've only scratched the surface.

From start to finish, it has taken just thirty days to complete what I set out to do, i.e., share my experience with Generative AI and ChatGPT.

Wrap-Up

I hope what you've seen and read has given you an insight into the power and possibilities of Generative AI. Writing the book has also been a learning experience for me, where my first draft was a loosely coupled connection of chats I'd created without much thought to how they cohesively fitted together.

I gave ChatGPT my table of contents and asked whether the information flowed. The response was positive with the caveat that without seeing any real content, it couldn't say with any certainty that it was achieving what I intended.

Throughout the book, I've referred to the value of human connection, and that's where reviews conducted by humans proved to be the difference. My "Triple-J" reviewers (Jaqui, Joanne, and John) looked at it from the perspective of someone who might be willing to pay for the content.

Their critique encouraged me to restructure that content, and hopefully, it has been worthwhile and contributed to your understanding of Generative AI and ChatGPT.

See What Else I'm Writing About

You can follow me here to see what else I'm writing about.

LinkedIn (Peter Stanford)

Blog (mentormeaus.com)

Email peter.stanford@mentormeaus.com

Figure 10 - Image generated by ChatGPT-4, 4th December 2023

DALL-E Image Description

A monochrome pencil drawing depicting a judge mediating a dispute about copyright for a book. The scene includes the judge, the author, the author's legal representative, the plaintiff, and the plaintiff's legal representative. The judge is shown in a position of authority, with the author and their legal representative on one side and the plaintiff and their representative on the other.

Appendix A – Plagiarism and ChatGPT

Earlier, I outlined the position taken by Microsoft and Open AI regarding their initiative called Copyright Shield. While this should give you comfort regarding anything you publish based on your chats with ChatGPT, don't rely on a legal defence if you haven't checked your content for plagiarism.

There are several free plagiarism checkers available. I use Grammarly because I'm familiar with it for spelling and grammar checking. To access the Grammarly plagiarism site, click the link https://tinyurl.com/ChatGPTPlagiarismMentorMeAus or scan the QR code below.

To demonstrate this, I copied the response from Bing into the Grammarly Plagiarism Checker. Here's the result.

It doesn't matter whether the content you check is from ChatGPT or Bing because ChatGPT powers Bing.

Here's a fun fact for you. I couldn't generate the QR code directly because the Grammarly page's URL was more than 250 characters. I converted that to a smaller URL using the TinyURL app: https://tinyurl.com/app. The name you see is the alias I created before generating the QR code. The alias must be no more than thirty characters.

NOTE: Most plagiarism checkers, Grammarly included, limit how much you can upload. The limits relate to the number of words and the overall file size. This book couldn't be checked in one attempt because the images exceeded the file size. Each plagiarism checker I tested allows you to copy and paste content into a text field before initiating the test. Breaking the text into smaller pieces doesn't negate the validity of the test.

NOTE: Turnitin allows much larger files to be uploaded. However, it requires a paid subscription.

The following transcript relates to a prompt outlined earlier in the book.

Response (Verbatim)

Thank you for your interest in ChatGPT and the issue of plagiarism. I will try to answer your question as best as I can.

First of all, you are correct that OpenAI (and Microsoft) have recently announced a new program called **Copyright Shield**, which aims to protect customers from legal claims surrounding copyright infringement[12]. According to OpenAI, this program will:

step in and defend our customers, and pay the costs incurred, if you face legal claims around copyright infringement.

This program applies to customers who use the **generally available** features of OpenAI's developer platform and the **business tier** of ChatGPT. However, there are some limitations and conditions that customers need to follow, such as:

- Customers must comply with OpenAI's Terms of Use, Usage Policies, and Sharing & Publication Policy.
- Customers must not use the Services for illegal or harmful purposes, or in a way that infringes, misappropriates, or violates any person's rights.
- Customers must not use output from the Services to develop models that compete with OpenAI, or represent that output from the Services was human-generated when it is not.
- Customers must notify OpenAI promptly of any legal claims they receive and cooperate with OpenAI in the definition of such claims.

Response (Verbatim)

- Customers must not admit liability, settle, or compromise any claims without OpenAI's consent.

You can find more details about the Copyright Shield program on OpenAI's website[1].

Secondly, you are partially correct in assuming that the training data for ChatGPT was in the public domain and therefore not likely to be challenged. ChatGPT is trained on a large corpus of conversational data from various sources, such as Reddit, Twitter, and OpenSubtitles. These sources are generally considered to be public domain, or at least have permissive licenses that allow for reuse and modification. However, there may be some exceptions or edge cases where some of the data may not be public domain, or may have more restrictive licenses. For example, some Reddit posts or comments may contain quotes or excerpts from copyrighted works, or some OpenSubtitles files may have non-commercial or attribution requirements.

Therefore, it is possible that some of the training data for ChatGPT may not be public domain, or may have different licenses than the rest of the data. However, this does not necessarily mean that the output from ChatGPT will be challenged or infringe any rights. This is because ChatGPT does not extract content verbatim from the training data, nor does it claim it as its own. ChatGPT uses the training data as a source of inspiration and knowledge, and generates new and original content based on the user's input and instructions. ChatGPT also does not retain or store any of the training data or the output it generates, so there is no risk of copying or distributing the data without permission.

Furthermore, OpenAI has stated that it will not claim any rights over the output generated by ChatGPT for the customers or their end users[3]. This means that the customers or their end users will own the rights to the output they generate using ChatGPT, and they can

Response (Verbatim)

use it for any purpose they want, as long as they comply with OpenAI's policies and the law. OpenAI also encourages the customers or their end users to give credit to ChatGPT when they share or publish the output, and to follow the best practices for responsible and ethical use of AI.

I hope this answers your question. If you have any more questions or feedback, please feel free to ask me. I am always happy to chat with you. 😊

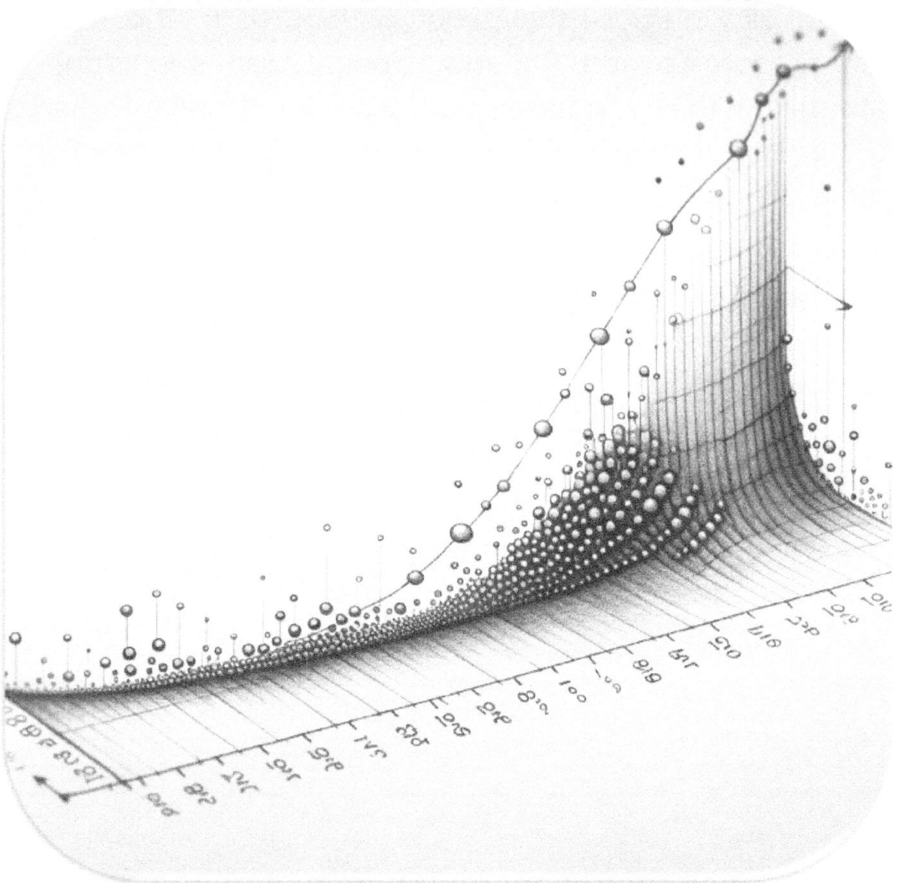

Figure 11 - Image generated by ChatGPT-4, 29th November 2023

DALL-E Image Description

A monochrome pencil drawing depicting a timeline of AI development as points on an exponential curve. The curve starts with a gradual incline, symbolising the early stages of AI, and then sharply rises to represent the rapid acceleration in the last ten years.

Appendix B – The AI (R)Evolution

If you really want to know about the history and development of AI, ask ChatGPT.

This is a practical book, and I encourage you to use ChatGPT to answer your questions. With that in mind, I asked ChatGPT to help me create a prompt you could use to explore the history of AI if you were interested. Here's the prompt.

Could you provide a detailed history of the development of Artificial Intelligence, starting from its early conceptual stages in the 1950s to late 2023? Focus on key milestones, including major theoretical advancements, technological breakthroughs, periods of stagnation or 'AI winters,' and notable 'AI summers' marked by rapid progress. Highlight influential figures, significant projects, and pivotal moments that have shaped AI's evolution. Please keep the response concise, targeting around 1,000 words to balance depth and brevity.

Will this give you everything you want? Possibly not, but you can change the prompt to ask for more or fewer words, change the perspective, or extend this prompt to include other information you'd like to know. I won't give you the complete response, but here's a summary of the major headings ChatGPT returned when I ran this query.

1. The Dawn of AI (1950s)
2. AI's Early Promise and Progress (1960s)
3. The First AI Winter and Reassessment (1970s)
4. Knowledge-Based Systems and Second AI Summer (1980s)
5. Machine Learning Emerges (1990s)
6. AI in the Mainstream and Commercial Success (2000s)
7. Deep Learning Revolution (Early 2010s)
8. AI Ethics and Advanced Applications (2020s)
9. Key Figures in AI Development
10. Technological Breakthroughs in AI
11. Understanding AI Winters and Summers
12. Significant AI Projects and Milestones
13. AI Today and Tomorrow: The Current State and Future Directions

Today, Generative AI stands not only as a testament to human ingenuity but also as a beacon, illuminating the potentialities of future innovations in a world where humans and machines collaborate more closely than ever.

Figure 12 - Image generated by ChatGPT-4, 28th November 2023

DALL-E Image Description

A monochrome pencil drawing depicting a lecturer standing at a whiteboard and facing a group of university students. The lecturer's body and face are directed towards the students, indicating direct engagement and interaction. The students, a mix of males and females, are attentively watching the lecturer, who is demystifying AI jargon with gestures towards both the whiteboard and the students.

Appendix C – Demystifying the Jargon

Let's get the jargon out of the way! I've explained what Prompts, Responses, Interactions, and Custom Instructions are in the Introduction, but the following terms are those you'll hear from time to time and are worth having a brief understanding of.

To learn more about them, create a prompt and ask ChatGPT. For example:

Please give me a 100–120-word description of Neural Networks.

A Neural Network is an advanced algorithmic structure in AI, modeled to simulate the human brain's functioning. It consists of interconnected nodes or neurons in multiple layers: an input layer, several hidden layers for data processing, and an output layer for results. The neurons perform computations and adjust their connections as data is fed through the network. This adjustment is a learning process, often supervised, where the network is trained with specific data to recognise patterns. Neural networks are adept at handling tasks like image recognition, speech processing, and predictive analysis, making them vital in complex AI applications.

Here are thirteen definitions provided by ChatGPT.

1. **Neural Network:** A system of algorithms that attempts to recognise underlying relationships in a data set through a process that mimics how the human brain operates.
2. **Machine Learning (ML):** A subset of AI where computers are trained to learn from data without being explicitly programmed for a task.
3. **Natural Language Processing (NLP):** A branch of AI that helps computers understand, interpret, and generate human language in a valuable way.
4. **Training Data:** The initial set of data used to help a machine or algorithm begin to "learn" and make predictions.
5. **Deep Learning:** A subset of ML, where neural networks with many layers (deep neural networks) process various data factors.
6. **Token:** In the context of NLP, a token is a sequence of characters, like a word or a sentence, that has some semantic value.
7. **Transformer Architecture:** A type of deep learning model architecture predominantly used in NLP tasks.
8. **Large Language Model (LLM):** An advanced type of neural network specifically trained on vast amounts of text. LLMs, like ChatGPT, can generate human-like text based on the patterns they've learned from their training data. Their large size, often containing billions of parameters, enables them to capture nuances and generate detailed responses.

9. **Parameters:** Elements within a neural network are adjusted during training to improve the model's accuracy. Think of them as dials or knobs that get fine-tuned so the model can make better predictions.
10. **Supervised Learning:** A type of machine learning where the model is trained on a labelled dataset, meaning the data is paired with the correct answer. The model makes predictions based on this training and adjusts its parameters to minimise errors.
11. **Unsupervised Learning:** Unlike supervised learning, the model is given unlabelled data in this method and must find patterns and relationships within the data independently.
12. **Generative AI:** A subset of AI that can create new content. It can be text (like ChatGPT), images, music, or videos. The AI learns from existing data and then generates something new based on that knowledge.
13. **Fine-Tuning:** After a model is pre-trained on a large dataset, it can be further trained (or "fine-tuned") on a smaller, specific dataset to make it more specialised for a particular task.

Figure 13 - Image generated by ChatGPT-4, 28th November 2023

DALL-E Image Description

A monochrome pencil drawing depicting an author sharing chats with clients via an online platform. The author is shown in front of a computer screen, sending a URL and a QR code to share the chats. The clients, a diverse group of males and females, are represented on the computer screen, receiving the information digitally.

Appendix D – Shared Chats

A feature of ChatGPT is the ability to share chats. The recipient can't edit these chats; however, the content can still be copied and used independently.

I've chosen to share several of my chats to help you understand how they were constructed and the thinking behind them, i.e., what did I start with, how or why did I extend that after each response, and when did I think I had sufficient quality information to form an opinion I was happy to share.

When and if you view these chats, you'll notice that I don't necessarily adhere to a single theme or topic. One of the key benefits I've found is letting my thoughts take me where they do. Part of a response may trigger something loosely related to the original prompt or topic, so I'll write the next prompt before I forget it.

I've shared ten of the 200+ chats I've had with ChatGPT. I hope they give you a sense of what you can ask, how to refine the question, and the information you get when you finish.

The links below contain a chat name and a precis generated by ChatGPT. There is also a URL to the chat and a QR code to look at it directly on your mobile device. All these links were verified before publishing.

I've removed any personally identifiable information from the precis. However, some may still exist in the chat responses because my custom instructions allow ChatGPT to refer to me by name or title. Anyway, you'll know who created them because my name's at the front of the book, and ChatGPT will often refer to me as Coach or Peter.

Some of these chats were created with ChatGPT-3.5. Those chats are not able to share images, while ChatGPT-4 can.

NOTE: As of 21st November 2023, chats containing images can not shared in ChatGPT-4.

Caution: When you share a chat, be conscious of whether that chat contains sensitive information. You may trust the person you're sharing the chat with, but nothing can stop them from sharing the URL with anyone else.

I've included ten chats in this list, curated from the more than two hundred I've worked on over the last eight months. To see what else I've been researching with ChatGPT, you'll find the links in the Wrap-Up section at the end of Chapter 6.

Chat Title

#35-The Psychology of Visionary Leaders

Precis

In a discussion with ChatGPT, key traits of visionary leaders were explored, highlighting long-term thinking, infectious passion, risk-taking, and adaptability. Visionaries possess charisma to rally people around their ideas and an innate curiosity to challenge the status quo. Identifying these leaders involves examining their track record, resilience, and consistent actions. An insightful exploration.

URL

https://chat.openai.com/share/b39def1a-d2a1-47df-a579-371346f03109

QR Code

Chat Title

#37-AI Thought Leader Strategy

Precis

A seasoned business and executive coach is pioneering the use of Generative AI in coaching. With a rich background in executive management, this coach is exploring how AI can refine coaching methods. The goal is to seamlessly integrate AI into their practice, offering clients advanced tools for personal growth and behavioural transformation.

URL

https://chat.openai.com/share/92332d7c-ac31-4200-89f9-7d2dec15e317

QR Code

Chat Title

#28-AI Balancing Fear with Facts

Precis

In a thought-provoking conversation with ChatGPT, Peter explored the various facets of AI, uncovering how biases in questioning can significantly influence the perception of technology. This discussion served as a potent reminder of the importance of maintaining a balanced approach in AI discourse, emphasising the need for nuanced understanding in this rapidly evolving field.

URL

https://chat.openai.com/share/918bf10e-32b6-4a73-84dc-1cb9f00edad5

QR Code

Chat Title

#13-Understanding Behavioral Characteristics

Precis

In a recent discussion, we explored the commonalities among professionals, emphasising that business and executive coaches, like others, experience fears and doubts. To highlight these shared challenges, we drew on research regarding imposter syndrome, burnout, emotional intelligence, and personal development. The key message? Coaches can embrace AI to enhance their practice, recognising it as a complement rather than a threat. #Coaching #AI #ProfessionalDevelopment

URL

https://chat.openai.com/share/6d0701c1-0ba2-4b5c-b706-bf35fe330677

QR Code

Chat Title

#23-AI and the Impact of Societal Change

Precis

In a compelling conversation with ChatGPT, we traversed the changing landscape of technology, employment, and social impact. From the Internet to breakthroughs in medical science like gene tech and early cancer detection, tech is not replacing us; it's redefining our roles and enhancing our lives. Yet, the question remains: How do we bridge the gap between tech optimists and those wary of change? The key is empathy, open dialogue, and education. Whether it's unionised resistance or government policies, understanding and trust are the bedrock on which we can build a more inclusive future. This is an insightful discussion for anyone navigating the intricacies of tech adoption, climate policy, and the future of work.

URL

https://chat.openai.com/share/40ccadc6-382e-48fb-b013-54948c99540d

QR Code

Chat Title

#18-AI Mentor Experiment

Precis

Peter is developing an innovative course that leverages Generative AI for self-exploration and goal-setting. This unique program teaches effective questioning techniques and how to interact with AI, like using personas for personalised advice. It's ideal for enhancing coaching and mentoring skills, requiring no prior AI knowledge. A game-changer in the coaching world is on the horizon!

URL

https://chat.openai.com/share/10fb6214-8c5a-4b83-9fc7-e7a23ad3ecd8

QR Code

Chat Title

#75-Export Capability for Chats

Precis

In a recent chat, Coach Peter explored the potential for integrating a targeted chat export capability in ChatGPT, aiming to manage his extensive conversations efficiently. He considered automating the process into a spreadsheet for easier tracking and content management. The discussion also covered the limitations regarding linking and modifying dates in ChatGPT, aligning with his professional needs in coaching and executive mentoring.

URL

https://chat.openai.com/share/58dc13ad-f582-4584-bc88-3485932ec712

QR Code

Chat Title

#105-Optimizing ChatGPT Within Limits

Precis

In this chat, the user explored writing a blog post about navigating ChatGPT's limitations ethically and effectively. The assistant provided strategies for leveraging the AI within its boundaries. Discussions also delved into the use of custom instructions for chat management, emphasising the need for clarity and the role of user feedback in refining AI interactions.

URL

https://chat.openai.com/share/8dd35788-bb7e-4ee0-b767-1880f8780ff9

QR Code

Chat Title

#113-Psychology of Musical Chairs

Precis

During a conversation on 22/11/2023, Coach, an experienced business and executive coach, asked about the psychological implications of the children's game "musical chairs." The inquiry focused on understanding what the game teaches children beyond its surface-level entertainment. The response highlighted various lessons embedded in the game, including competition, decision-making, strategy, resilience, physical coordination, social interaction, and adherence to rules. This analysis aligns with the Coach's interest in blending traditional coaching with Generative AI to provide insightful perspectives, particularly in the context of behavioural change and self-awareness. The response aimed to provide a nuanced understanding of a seemingly simple children's game, highlighting its potential developmental benefits and alignment with broader coaching and leadership principles.

URL

https://chat.openai.com/share/b1979da5-6460-4e1e-94b9-a1ec088796d7

QR Code

Chat Title

#120-Ask Better Questions

Precis

This conversation focused on integrating AI into business, personal development, and coaching practice. Key topics included the NDIS, the Rugby World Cup, the importance of understanding personal strengths in coaching, AI's role in business writing and societal impact. Discussions about specific AI tools for picture creation and PDF examination and the psychology behind individual responses to AI-induced job changes were also prominent. These insights aimed to contribute to your book, *The Little Book of Chats*, exploring AI's practical applications and influence on human behaviour and professional development, aligning with your objective of combining traditional coaching techniques with Generative AI.

URL

https://chat.openai.com/share/f24de116-e4e4-49eb-8234-9386d676e55e

QR Code

Acknowledgements

All authors say it, and we all mean it. No matter how much time and energy we invest in writing, someone else will view it far more dispassionately and provide important feedback to ascertain whether the content will attract interest.

I've been fortunate to have my Triple-J network of reviewers: Jaqui Lane (publisher), Joanne Curry, and John Forrest (colleagues and friends). Reviewing 160+ pages isn't a simple task, and I thank them for their time and insight into whether this book hit or missed the mark.

A combination of ChatGPT and Grammarly should have caught any errors, and lastly, so should I. If errors persist, the responsibility is mine alone.

NOTE: Other than the few references shown in the References section and specific references generated by ChatGPT included in the text in Chapter 5, all other content has been part of the Generative AI process that is ChatGPT.

References

How Fast is Coaching Changing, retrieved December 2, 2023, from https://clutterbuck-cmi.com/blogs/how-fast-is-coaching-changing/

Jeffers, S. (1987). *Feel the Fear and Do It Anyway*. Ballantine.

Mentoring in the AI World, retrieved December 2, 2023, from https://chronus.com/blog/mentoring-in-the-ai-world.

Passmore, J. & Woodward, W. (2023) Coaching education: Wake up to the new digital and AI coaching revolution! International Coaching Psychology Review, 18(1). 58-72

Index